# Advances in Soil Science

# Advances in Soil Science

B.A. Stewart, Editor

# Advances in Soil Science

## Volume 2

Edited by B.A. Stewart

With Contributions by
Elaine Bragg, J.M. Lynch, Konrad Mengel, Calvin W. Rose,
and Koji Wada

With 54 Illustrations

Springer-Verlag
New York Berlin Heidelberg Tokyo

B.A. Stewart
USDA Conservation & Production Research Laboratory
Bushland, Texas 79012 U.S.A.

ISBN-13: 978-1-4612-9558-7          e-ISBN-13: 978-1-4612-5088-3
DOI: 10.1007/978-1-4612-5088-3

Typeset by Ampersand Publisher Services, Inc., Rutland, Vermont.

9 8 7 6 5 4 3 2 1

# Preface

The world population in 1930 was 2 billion. It reached 3 billion in 1960, stands at 4.6 billion today, and is expected to reach 6 billion by the end of the century. The food and fiber needs of such a rapidly increasing population are enormous. One of the most basic resources, perhaps the most basic of all, for meeting these needs is the soil. There is an urgent need to improve and protect this resource on which the future of mankind directly depends. We must not only learn how to use the soil to furnish our immediate needs, but also ensure that the ability of the soil to sustain food production in the future is unimpaired. This is indeed a mammoth task; a 1977 United Nations survey reported that almost one-fifth of the world's cropland is now being steadily degraded. The diversity of soil makes it necessary for research to be conducted in many locations. There are basic principles, however, that are universal. This, *Advances in Soil Sciences*, presents clear and concise reviews in all areas of soil science for everyone interested in this basic resource and man's influence on it.

The purpose of the series is to provide a forum for leading scientists to analyze and summarize the available scientific information on a subject, assessing its importance and identifying additional research needs. But most importantly, the contributors will develop principles that have practical applications to both developing and developed agricultures. It is not the purpose of the series to report new research findings because there are many excellent scientific journals for that need. Communications in scientific journals, however, are generally restricted to short and technical presentations. Therefore, *Advances in Soil Science* fills a gap between the scientific journals and the comprehensive reference books in which scientists can delve in depth on a particular subject relating to soil science.

The ultimate aim of the series is to stimulate action—action to determine where there are arable soils, action to develop technology for more efficient crop production on these soils, action to reduce the risk of degrading these soil resources, and action to determine on which soils our research efforts should be concentrated. Without such action, the task of producing adequate food in the future may simply be too great. By the time the world gets reasonably close to population stability, demand for food and other agricultural products could be three times present levels.

There are many audiences to reach. While intended primarily for scientists and students of soil science, this series also provides technical information for anyone interested in our natural resources and man's influence on these resources. The reviews are written by leading scientists from many countries, and therefore provide the reader with information from a wide array of conditions. Such information is particularly useful to professionals working in areas with developing agricultures because the reviews summarize and assess the significance of the technical literature.

B.A. Stewart

# Contents

# Contributors

ELAINE BRAGG, AFRC Letcombe Laboratory, Wantage, Oxon, OX12 9JT
England
J.M. LYNCH, Glasshouse Crops Research Institute, Rustington, Worthing
Road, Littlehampton, West Sussex, BN17 6LP England
KONRAD MENGEL, Institut für Pflanzenernaehrung, D-6300, Sudanlage 6,
Federal Republic of Germany
CALVIN W. ROSE, Griffith University, Nathan, Queensland 4111, Australia
KOJI WADA, Faculty of Agriculture, Kyushu University 46, Fukuoka 812,
Japan

# Developments in Soil Erosion and Deposition Models

## Calvin W. Rose*

*(continued)*

*Griffith University, Nathan, Queensland 4111, Australia.

© 1985 by Springer-Verlag New York
Advances in Soil Science, Volume 2

# I. Introduction

Research on water-induced soil erosion illustrates two useful approaches to mathematical modeling. The first, and often the prior type of model, is designed to produce order from a possibly large body of data about the system of interest and to derive conclusions from and summaries of that data base. This type of model may be informed by physical or other appropriate insight into system structure or behavior but depends strongly on the data and on mathematical or statistical modeling concepts.

The second type of model has its major focus in attempting to describe mathematically what is perceived as the major significant features of the structure of the system of interest and the processes operating on and within it. This type of model can be outlined even without data, even though it is common experience that data will be required to determine the value of parameters defined in this process-oriented type of model.

For complex systems, development of this second type of model commonly is undertaken following clarification of major features of system behavior by work resulting in the first type of model.

The Universal Soil Loss Equation (or USLE) of Wischmeier and Smith (1978) is an example of the first type of model outlined above. The USLE is based on a vast body of field data collected in the U.S. midwest on soil loss from plots of standard length but varied slope, land management, and rainfall characteristics. A statistical summary greatly enhances the usefulness of this data base.

There are difficulties for a country that is unable to repeat for itself the considerable research effort on which the USLE is based but wishes to use this approach to conservation planning, which has been extensively tested in the U.S. in particular. As reviewed for tropical countries by El-Swaify *et al.* (1982), the difficulty is that empirically derived "values for individual parameters in the USLE as they now exist, are restricted in validity to the mainland United States and can be extrapolated for use in the tropics only after experimental confirmation or modification." Built-in correlations between rainfall and runoff specific to the data set from which the USLE is derived is an example of how universality in application is restricted.

Despite such difficulties, the existing experience in use of the USLE, its good documentation, and its identification of factors important in erosion and its control combine to sustain a case for research designed to do the testing and modification required to support adoption elsewhere of the USLE methodology.

However, the recognition of such difficulties has encouraged interest in the second type of model referred to above. This interest is because basic processes are universal; thus, if success can be achieved in mathematically describing basic processes, the problems of transferability of such a model are much less than in a statistical data-based type of model first described. This is not to say specific local values for parameters are unnecessary.

## A. Scope of This Review

This review is limited to models of soil erosion by water and the subsequent deposition of the soil, and the models are of the second type described above, i.e., intended to describe the processes involved. A selective review will also be given of research on the processes involved in soil erosion and deposition.

Foster (1982) has recently published a substantial review of soil erosion models by himself and colleagues in related and unrelated institutions in the U.S., with Foster and Lane (1981) providing a complementary overview. Because these reviews are recent and because that by Foster (1982) is rather complete concerning the major relevant research in the U.S., this review emphasizes research and model development in Australia. Care is taken, however, to show how such models can be used to give an alternative model representation of processes to that reviewed by Foster (1982). Indeed, a basic and as direct as possible comparison between the alternative model approaches is made. Examples of application of the models presented in more detail are given, using data from a number of countries and a variety of environments.

## B. Basic Approach Adopted to Sediment Transport

Sediment flux ($q_s$) is defined as the oven-dry mass of sediment flowing across unit width perpendicular to the direction of the flux in unit time. By definition of terms, it is related to the volumetric water flux $q$ (also per unit width) and to effective average sediment concentration ($c$, expressed as oven-dry mass of sediment per unit volume of suspension) by the equation[1]

$$q_s = qc \qquad (\text{kg } \text{m}^{-1}\text{s}^{-1}). \qquad [1]$$

Describing $q$ in equation [1] involves hydrologic theory. The magnitude of $c$ arises from the balance between processes tending to add sediment to overland flow and the continuous removal of sediment by deposition. Deposition is the process whereby sediment settles out of the suspension because of its positive immersed weight. In contrast to Foster (1982), for example, the term "deposition" is therefore used to describe a process that continues while sediment with positive immersed weight is in suspension. Considerable difficulty is avoided if this use of the term "deposition" to describe a continuous process is distinguished from its common alternative use (Foster, 1982) in the sense of a net and possibly intermittent process.

Recent developments in models to represent the water flux $q$ in equation [1] will be reviewed first before turning to research on sediment-related processes. The question of overland flow can be considered largely ignoring its potential role in sediment transport. Sediment transport cannot be considered in ignorance of overland flow.

## II. Hydrologic Models

The objective in a predictive hydrologic model of relevance here is to be able to predict the time and space varying characteristics of overland flow from a measurement of rainfall rate ($P$). This requires a knowledge of an effective mean infiltration rate ($I$) for the management unit of land in question and how this varies with time ($t$).

A significant difficulty in using traditional infiltration theory is the large spatial variability in soil properties, such as hydraulic conductivity and sorptivity, on which infiltration rate depends. Such variation can be measured, but it is still a resource-consuming and difficult task to obtain an effective mean, even for a small subcatchment, and the results obtained are sometimes disappointing.

The reason for disappointing results may be not only the difficulty in coping with spatial variation, but, at least in regions of higher rainfall

---

*A list of major symbols is given in the Appendix (pages 58–59).

rates, the effect of the resultant structurally damaged surface soil layer in modifying infiltration rate.

Such difficulties could be avoided if it were possible to infer $I(t)$ from measurements of rainfall $P(t)$ and runoff rate for the management unit or subcatchment whose mean or overall infiltration characteristics are sought. The impediment to this procedure has been the lack of analytical solution for overland flow to all but a step function in excess rain $(R)$ (Woolhiser and Liggett, 1967; Lane and Shirley, 1982), where $R$ is defined by

$$R = P - I,$$

so that

$$I = P - R, \qquad [2]$$

an equation in which all terms are assumed variable in time, but not in space at any time.

Numerical solutions of the equations appropriate for overland flow of mass and momentum conservation (normally simplified by the kinematic flow approximation) may be obtained. However, these numerical procedures are described for determining runoff assuming $R$ to be known, whereas the practical case addressed here is the reverse, so that with $R$ determined $I$ can be calculated from equation [2]. Let

$$Q = \text{runoff rate per unit land area.}$$

The impediment provided by the lack of analytical solutions between $Q$ and $R$ for overland flow for the general case has recently been overcome. General exact and approximate analytic solutions are now available and will be briefly described.

## A. The Exact Solution for Overland Flow of Parlange *et al.* (1981)

Parlange *et al.* (1981) have provided an exact general solution for a planar land element. Conservation of mass is given by

$$\frac{\partial q}{\partial x} + \frac{\partial D}{\partial t} = R, \qquad [3]$$

where $x$ is distance down the plane from where the depth of flow $(D)$ is zero (which provides the boundary condition for the solution).

The kinematic flow approximation can be expressed in the form

$$q = K_1 D^m, \qquad [4]$$

where

$$K_1 = S^{1/2}/n, \qquad [5]$$

with $S$ the slope of the plane (the sine of the slope angle), $n$ Manning's roughness coefficient for the plane, and $m \sim 5/3$ for turbulent flow.

The solutions of equations [3] and [4] (for any $R$), with the initial condition $D = 0$ at $t \leq 0$, are

$$D(t) = \int_0^t R d\bar{t} - \int_0^{t_0} R d\bar{t}, \qquad [6]$$

$$x = K_1 m \int_{t_0}^t \left[ \int_{t_0}^{\bar{t}} R(\bar{\bar{t}}) d\bar{\bar{t}} \right]^{m-1} dt, \qquad [7]$$

where $t_0$ and so $\int_0^{t_0} R dt$ are (variable) constants of integration, and $\bar{t}$ and $\bar{\bar{t}}$ are variables of integration. There are two branches to the solution that join at $x(t) \equiv x_p$ given by equation [7] with $t_0 = 0$. The junction $x_p$ moves down-slope with a velocity proportional to the flow velocity at $x_p$. The time at which $x_p$ reaches the end of the plane $x = L$ is called the transition time, $t_\tau$, or alternatively the time of concentration.

For the solution branch downstream of $x_p$, Parlange et al. (1981) show that $t_0 = 0$, whereas upstream of $x_p$, $t_0 > 0$, and its value is a function of $x$. From equation [6], if $t_0 = 0$ (i.e., $x > x_p$), then $D$ is a function of time only, and not $x$. Thus, down-slope of $x_p$, depth $D$ is constant, but $D$ varies with $x$ (as well as $t$) upstream of $x_p$ (see Figure 1). Curve E2 in Figure 1 shows the situation where $x_p > L$, $t > t_\tau$, and $t_0 > 0$.

The existence of different solutions up-slope and down-slope of $x_p$ in this exact theory adds to complexity, and the use of numerical methods required to evaluate integrals adds to computational cost. These complexities remain when the above exact theory is adapted to cope with the common practical situation of inferring $R$ from measurements of $P$ and $Q$, thus yielding $I$ (equation [2]).

Hence, it has proven useful in practice to use an approximate analytic solution to the above problem, which also satisfies kinematic flow assumptions, but makes other approximations as described below.

## B. The Approximate Analytic Solution for Overland Flow of Rose et al. (1983b).

Depth of overland flow ($D$), even for a relatively flat agricultural land surface, is clearly a somewhat stochastic variable. Thus, approximate theory in which there is some generally modest error in $D$, but in which the water flux $q(L)$ at $x = L$, the exit from plane, is the same as for the exact theory and experimental measurement, should be useful if it is substantially simpler to apply than the exact theory described above.

If a steady-state flow situation exists, then

$$Q = R, \qquad [8]$$

where by definition of $Q$, quite generally,

$$Q = q(L)/L. \qquad [9]$$

For a steady state, mass conservation requires that

$$q(x) = Rx, \qquad [10]$$

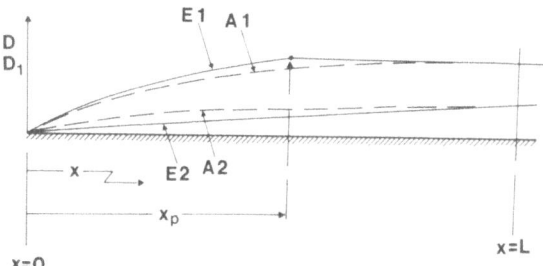

**Figure 1.** Variation in overland flow depth with distance $(x)$ down a plane of length $L$, from exact $(D)$ and approximate $(D_1)$ solutions of equation [3]. Exact $(E)$ and approximate $(A)$ profiles are shown at an earlier time $(E1, A1)$, where the distance $(x_p)$ to the junction of the two branches of the exact solution is less than $L$, and at a later time $(E2, A2)$, when $x_p > L$.

and so from equation [4] the depth of flow at steady state $(D_s)$ at $x$ is given by

$$D_s = (Rx/K_1)^{1/m}. \qquad [11]$$

The approximation now made is this: Even in non-steady–state flow situations the algebraic (and thus geometrical) form of the relation between water depth and $x$ will be taken to be the same as in the steady state (equation [11]), but with $R$ replaced by $Q$. Temporarily denoting the approximate water depth that follows from this assumption by $D_1$, then

$$D_1 = (Qx/K_1)^{1/m}. \qquad [12]$$

The approximation given by $D_1$ to $D$ is illustrated by comparing curves A (approximate) to E (exact) in Figure 1, for an earlier (1) and later (2) time in a runoff event.

It may be noted that at $x = L$

$$D_1 = (QL/K_1)^{1/m}$$
$$= (q(L)/K_1)^{1/m} \text{ from equation [9]}$$
$$= D \qquad \text{from equation [4],}$$

so that the approximation $D_1$ to $D$ given by equation [12] ensures exact agreement in depth (and thus in water flux or runoff) at $x = L$, as is also illustrated in Figure 1. The cost of this agreement in runoff and great simplicity in form of the profile shape (equation [12]) is that mass of water on the plane is not conserved at all times during the runoff event, although it is conserved for the event as a whole.

Let us now determine the relationship between $R$ and $Q$ yielded by this approximation. Using the initial boundary conditions that

$$R = 0, t \leq 0, \text{ and}$$

$$q(0) = 0 \text{ for all } t,$$

equation [3] can be integrated from $x = 0$ to $L$, giving

$$q(L) = RL - \int_0^L (\partial D/\partial t)dx \qquad [13]$$

Using the approximation $D_1$ (equation [12] for $D$ in equation [13], and using equation [9],

$$R \doteq Q + (1/L)\int_0^L (\partial D_1/\partial t)dx. \qquad [14]$$

Substituting for $D_1$ from equation [12] in [14] yields:

$$R = Q + [m/(m + 1)](L/K_1)^{1/m}(dQ^{1/m}/dt)$$

$$= Q + K_2(dQ/dt), \qquad [15]$$

where

$$K_2 = [1/(m + 1)](LQ^{1-m}/K_1)^{1/m}. \qquad [16]$$

Equation [15] allows $R$ to be calculated (approximately) from $Q$ (given $L$, $K$, and $m$). If the approximation to $R$ given by equation [15] is denoted $R_a$, then the form of the relationship between $R_a$ and $Q$ is illustrated in Figure 2 for an assumed simple form for $P$.

Note that it follows from equation [15] that $R_a = Q$ at the maximum in $Q$ when $(dQ/dt) = 0$. Also if $(dQ/dt)$ is positive, then from equation [15] $R_a > Q$, whereas if $(dQ/dt)$ is negative, then $R_a < Q$ (Figure 2). Figure 2 also illustrates that $I$ becomes less than $P$ when $R > 0$ following the commencement of ponding (equation [2]). Note that $R$ can be negative if, as is shown in Figure 2 towards the end of the rainfall event, $I > P$.

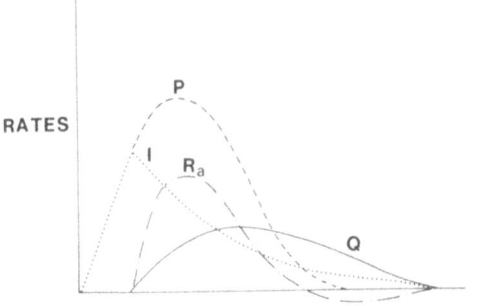

RATES

TIME FROM RAINFALL COMMENCEMENT

**Figure 2.** Simplified time variation in rainfall rate ($P$), infiltration rate ($I$), rate of runoff per unit area of plane ($Q$), and $R_a$, which is the approximate analytic solution for the excess rainfall rate ($R$).

## C. Use of Solution Given in Subsection B to Determine Infiltration Rate ($I$)

With $R$ from equation [15], then $I$ is directly available from equation [2] as an effective spatially averaged value for the entire subcatchment that contributed to runoff per unit area $Q$.

Rose et al. (1983b) illustrate this method of determining $I$ and consider the magnitude of error possible as a result of the approximation made in the theory to achieve a simple analytic solution. The error so introduced in calculating $R$ is chiefly one of timing, not of magnitude. Thus, the consequences of error resulting from the approximation is of little or no consequence in terms of use of the theory of Rose et al. (1983b) as the hydrologic component of an erosion/deposition model.

The consequences of an error in timing introduced into $R$ can be more significant in terms of estimating $I$, however, since as equation [2] indicates, the value of $I$ depends directly on differences in the time-variable quantities $P$ and $R$.

Error can be introduced into $I$ for other reasons than the analytic approximation described in $B$. A common cause of error in the field is that uniform sheet flow of the type assumed does not take place. For plots tilled either cross-slope or up- and down-slope, for example, Loch and Donnollan (1983a) observed considerable variability in flow depth at the same distance down a field plot subject to simulated rainfall because of preferred drainage paths. Another cause of error is discrepancy in clock times between rainfall and runoff rate recorders, if these have independent clocks as has been common in the past.

Rose et al. (in press) illustrate how error introduced into $I$ as a result of any of the causes described in the paragraph above can be corrected for, at least approximately, making use of known general characteristics of $I(t)$, namely that it should not be negative and in general will reach a maximum value and then decline with time (e.g., Figure 2).

Runoff from a plane land surface may be measured directly or after collection in a channel. Such a collection channel is formed by that common upland soil conservation structure, the contour bank or graded channel terrace. Here, runoff collected in the channel flows approximately cross-slope at much reduced velocity (Figure 11). The flow is conveniently measured as it leaves the channel (for safe disposal in a grassed waterway, for example) using a hydraulic flume. The calculations, described in Rose et al. (in press) and aided by a computer program, GNFIL, involve application of equation [15] both to the cultivated plane and the graded channel.

The Department of Primary Industries, Queensland, has collected data of the type described in the previous paragraph. Figure 3 shows results from Rose et al. (in press) for $I$, the infiltration rate into the cultivated plane, calculated with this program. Rainfall rate is also shown.

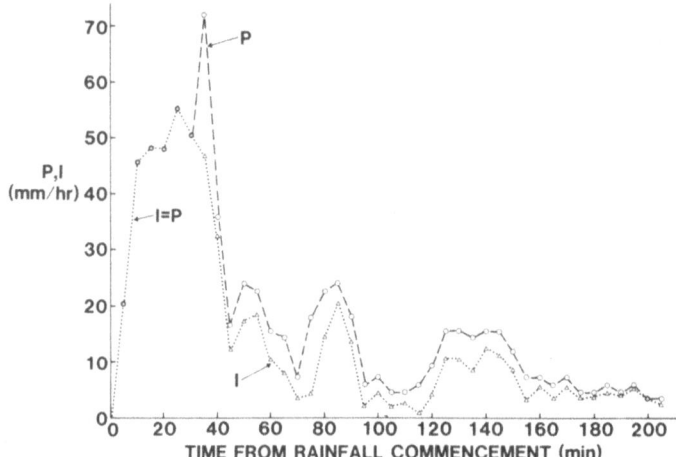

**Figure 3.** Data from Rose *et al.* (in press) showing the spatially averaged infiltration rate (*I*) for a cultivated field in the Darling Downs, Queensland, calculated from measurements of rainfall rate (*P*) and runoff rate using the approximate analytic hydrologic theory of Rose *et al.* (1983b).

The ability to utilize natural rainfall and standard measurement methods for *P* and *Q* confers a degree of practicality and efficiency to this method of determining *I*; the automatic spatial averaging aids its practical utility.

Figure 3 illustrates some complexity in time variation in *I* when *P* varies in the somewhat erratic fashion typical of natural rainfall. However, when *I* falls below *P*, the general decline in *I* with time expected on the basis of traditional infiltration theory can be recognized.

Using a constant value of *P* via sprinkling infiltrometer, Schroeder *et al.* (1982) used the value of *I* inferred from cumulative runoff from 1.4 m² plots, together with a modification of the Green and Ampt infiltration equation, to infer the saturated hydraulic conductivity of a field soil (assumed constant with depth in the theory). Because of the small plot size, it was not necessary to use hydrologic theory. This type of analysis can also be done with data such as illustrated in Figure 3, if assumptions behind the infiltration theory are adequately satisfied.

Infiltration rate is known to depend on many factors such as soil type, surface management, and antecedent water content. However, with data on *I* obtained as described above for a range of management or antecedent water content, it is then possible to develop empirical infiltration equations informed by the rich field of deterministic theory developed to describe the infiltration process under idealized conditions (reviewed, for example, in Hillel, 1980, and used by Schroeder *et al.* (1982) as mentioned above).

Such empirical equations, which allow the prediction of *I* from measured

$P$ (given other relevant information), are by nature somewhat location and management specific.

The hydrologic literature has considered the approximate representation of land surfaces more complex than the plane or plane plus channel considered above. V-shaped watersheds and the representation of a sloping surface by a cascade of planes have been studied, and the effect on overland flow simulation of simplification of watershed geometry considered (e.g., Lane and Woolhiser, 1977). An interaction between the degree of geometrical simplification and optimal values of hydraulic roughness parameters was noticed.

Overland flow on converging and diverging surfaces has also received analysis (e.g., Campbell *et al.*, 1984). The behavior of numerical procedures can be checked and their performance improved by using similarity solutions of the analytic equation for flow on such surfaces, even though general analytic solutions are not yet available (Campbell and Parlange, 1984).

Assumptions similar to those illustrated for a plane surface in subsection B above can be made to yield approximate analytic solutions for overland flow on conical (diverging flow) and inverted cone (converging flow) surfaces. Such solutions are of the same form as equation [15] except that the coefficient corresponding to $K_2$ is different from that given by equation [16].

## III. Controlled Research on Soil Erosion and Deposition Processes

Models of soil erosion and deposition are ways of expressing our concepts of the processes involved. Thus, it is important to review research designed to clarify such processes.

An understanding of processes involved in erosion, deposition, and transport of sediment has been significantly aided by studies with some control over variables. Laboratory flume studies and field studies using simulated rainfall are examples of the type of controlled research that will be reviewed in this section. The test of ideas suggested by such controlled-factor research still lies in field measurements under natural rainfall, and data of this kind with be reviewed in later sections.

This review of controlled-factor research on soil erosion and deposition will again, for reasons given in section IA, concentrate on research carried out in Australia. For reasons of space this review will be further restricted to the research of one laboratory and one set of field studies, representing work by the Commonwealth Scientific and Research Organization (CSIRO) and a State Government Department, respectively.

## A. Process-Oriented Research by CSIRO Division of Soils, Canberra, A.C.T.

The most developed laboratory facility in Australia to investigate sediment transport is in the CSIRO Division of Soils in Canberra. The work of this laboratory arose from field studies in geomorphology and fluvial processes in particular. Geomorphologists are heavily involved in these laboratory studies, and the reporting of these studies includes close linkage to field observation.

Walker *et al.* (1977) have described this rainfall simulation and tilting flume facility (of dimensions 3 × 0.6 m). Their research on the processes at work in sediment transport in shallow overland flows, with and without rainfall, will be reviewed in terms of the process consequences of gradually increased slope and water discharge.

Figure 4 illustrates the general finding that at very low slopes, and therefore low overland flow velocities, removal or detachment of sediment from the soil surface into the shallow water layer is due to rainfall detachment alone. For a constant water flux $q$ (as in the experiments of Figure 4), sediment flux $q_s$ will be proportional to sediment concentration $c$ (equation [1]). Thus, figure 4 illustrates that $c$ is insensitive to slope ($S$, the sine of the slope angle) up to $S \sim 0.03$. This insensitivity would be expected since $c$ would depend on the balance between rates of rainfall detachment

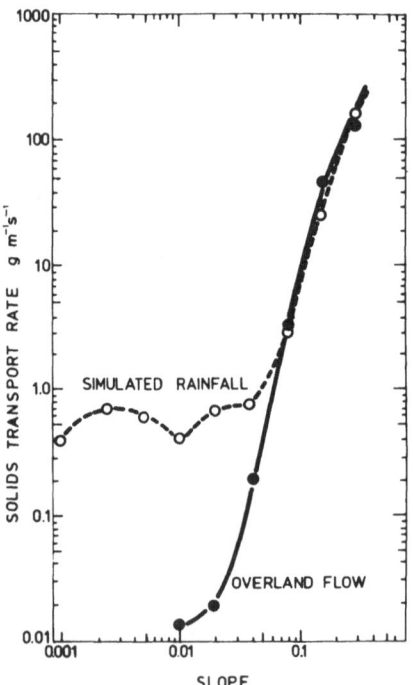

**Figure 4.** Sediment flux ($q_s$) measured on a laboratory flume at various slopes ($S$). The dashed line indicates the relationship between $q_s$ and $S$ where all the water flux $q$ was provided by simulated rainfall. The solid line indicates the same relationship with the same water flux $q$, but now provided entirely as overland flow (Moss, 1979).

and deposition, neither process being sensitive to slope. This regime where rainfall is the only agent of sediment detachment and surface flow acts simply to transport and not generate sediment, has been termed "rain-flow transportation" by Moss *et al.* (1979).

At such low slopes sediment is splashed into the air in random directions, unlike the situation with sloping surfaces where preferential downhill splash can occur. Normally, such aerial splash transport makes only a minor contribution to total sediment transport (Walker *et al.*, 1978).

If the water layer on shallow slopes is of sufficient depth to protect the soil surface from raindrop action, then sediment concentrations and rain-flow transport would be negligible. Moss and Green (1983) found rain-flow transportation rates to rise from zero at zero flow depth to near maximum values at flow depths corresponding to 2 to 3 drop diameters. Such transport decreased rapidly with further increase in water depth, being one to two orders of magnitude lower than the maximum for a water depth of 5 drop diameters for the 200 μm quartz sand used in these experiments.

As bed slope increased beyond 0.01 (Figure 4), the ability of overland flow alone to entrain and transport sediment rose rapidly, equaling that of rain-flow transport as $S$ reached ~0.05. For $S > 0.05$, entrainment by the surface flow (or runoff detachment) became the dominant mechanism contributing to sediment transport. This conclusion is indicated by the negligible difference in Figure 4 between transport whether rainfall or overland flow provided the discharge of water. The discharge was identical for either form of water application. (Hereafter I will use the term "entrainment" for runoff detachment, the term being commonly used, especially in the geomorphological literature.)

Though the overall character of the relationships in Figure 4 is expected to be general, the particular critical slopes are not universal. The results of Figure 4, obtained with a poorly sorted sand, when compared with those of Singer and Walker (1983) who used a fine sandy loam yellow podzolic (Albaqualf), indicate the importance of the soil material. For the same slope (0.09) and similar overland flow and rainfall rates (100 mm h$^{-1}$), rate of soil loss was an order of magnitude higher for the podzolic if all water was applied as rainfall compared with all as overland flow, whereas there was little difference for the sand.

As slope and discharge increase, the uniform or nearly flat beds characteristic of rain-flow transport tend to be replaced by channels because of instability and secondary flow effects (Moss *et al.*, 1982). However, even at slopes or discharges when rain-flow no longer dominates sediment transport, rainfall is observed to play another role of stabilizing sheet flow in situations where otherwise rills or small channels would develop (Moss *et al.*, 1979). An implication of this is that rills will become more pronounced during runoff following cessation of rainfall. Thus, after a rainfall event, rill systems may be left in a more strongly developed and pronounced form than was typical for the event as a whole. With the same discharge maintained, but the proportion of rainfall to surface flow varied, Moss *et al.* (1979) found that the slope required to develop micro-channels increased with the proportion of rainfall. The spatial uniformity of turbulence generated by rainfall compared with bed-friction–generated turbulence was thought to be responsible for this observation.

The structures characteristic of deeper flows, such as ripples and dunes, are greatly suppressed in shallower overland flow (Moss and Walker, 1978). However, in overland flow small channels are a typical feature encountered as slope or discharge increases, and transport mechanisms in such channels have similarities to those in streams and rivers (Moss and Walker, 1978; Moss *et al.*, 1982).

Moss and Walker (1978) suggest that the typical tendency of loess to form incised rills that expand and capture each other to form gullies is related to the general absence of particles greater than silt size. With larger sized particles available, rills tend to be lined with their own deposits, especially in areas of net deposition—a situation referred to as "bed-load amelioration."

Transport of loose, sandy detritus in the absence of rainfall was investigated by Moss *et al.* (1980, 1982). At low slopes and flow rates, armoring of the bed and saltation of individual sediment particles were observable. As flow rates increase, interparticle collisions become so numerous that saltation of individual particles can no longer be observed, though it is presumed that the basic process of saltation still continues. This more developed state of sediment transport is described as "rheologic flow" (Moss, 1972). Bagnold (1956) called the same phenomena "ballistic dispersion"; it is this type of sediment transport that dominates in the field when erosion losses from single events are noticeable.

The distinction between suspended load and bed load commonly made in streams can be observed in lower velocity shallow overland flows. Bed-load transport can be composed of sediment transported by saltation and rolling of the larger particles forming the contact load. Moss *et al.* (1979) found that in situations where they could be distinguished, the suspended, saltating, and contact (rolling) loads normally can be associated with particular sediment size ranges, with suspended load being <30 μm in size. A ranking in speed of transport was inferred, with suspended load > contact load > saltation load being typical.

Moss and Walker (1978) argue that in overland flow, apart from temporary surges, the rate of suspended and bed-load transport must be substantially correlated, even if they are separable. Moss (1972) observed the depth of "rheologic layers" to increase with flow rate until they filled the entire depth of flow—the situation during most of a field erosion event of significance.

Moss *et al.* (1980) found that there were characteristic sediment concentrations associated with the formation of rheologic layers for sediment of a particular size, being about 4 kg m$^{-3}$ for fine and medium sand. However, such characteristic concentrations were found to increase with particle size, so that the concept "transport capacity" is certainly not a characteristic of flow alone.

The slope dependence of sediment transport was found to vary with the type of transport mechanism. With sediment transported dominantly by the saltation of individual particles (as occurs at lower slopes and discharges), transport rate increased rapidly with slope.

For "rheologic flow" a less rapid increase of sediment transport with slope was observed than was the case for saltation-dominated flow.

For experiments characterized by rheologic transport and channels in the bed, sediment transport rate was found to be linearly related to stream power as shown in Figure 5 (Moss et al., 1980). Defined as the product of bed shear stress and flow velocity, stream power is proportional to the product of bed slope and discharge for overland flow.

From Figure 5, neglecting a small threshold effect,

$$q_s = kSq, \qquad [17]$$

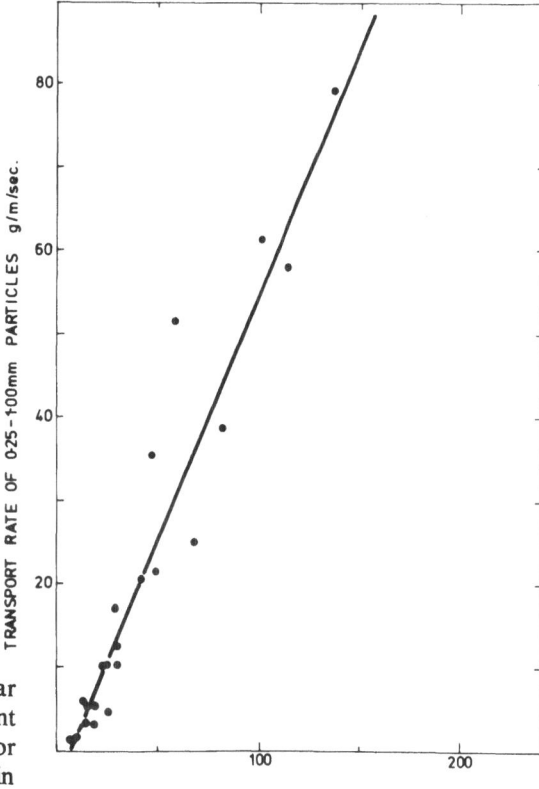

**Figure 5.** Illustrating the linear relationship between sediment flux ($q_s$) and stream power for rheologic flow with channels in the bed on a laboratory flume (Moss et al., 1980).

where $k$ is a dimensional constant. From equation [1], the result in equation [17] can be alternatively expressed as

$$c = 595S \qquad (\text{kg m}^{-3}) \qquad\qquad [18]$$

where $k = 595$ kg m$^{-3}$, a number specific to the experiment.

Moss *et al.* (1980) investigated rates of soil loss from beds with roughness amplitudes on the order of 100 mm. Following an initial surge with higher values, soil loss rates were similar to those from smooth-bed equivalents. Rills and channels, as well as cultivation, introduce spatial variability. However, these results on the effect of "cultivation," and those summarized in equation [18], provide encouragement that at least to a first approximation, sediment flux can be interpreted in terms of spatially uniform models that lack the detailed geometry of the actual complex surface structures transporting sediment.

### B. Process-Oriented Research by Queensland Department of Primary Industries

Since work by Freebairn *et al.* will be referred to later, this section will be restricted to the work of Loch and Donnollan (1983a, 1983b), who studied two clay soils (a Pellustert and an Oxic Haplustalf) using a field rainfall simulator. Slope (4%) and rainfall rate (95 mm hr$^{-1}$) were constant, and use of plots of different length resulted in a substantial range of plot discharge.

Sediment concentration measured at exit from the plot was found to have one of two values for any particular soil. A lower value was measured at lower plot discharges, and beyond a certain critical discharge, corresponding to a stream power of 0.05 W m$^{-2}$ for both soils investigated, the sediment concentration increased rather suddenly (Figure 6). The values of these two concentrations differed for the two soils. Sediment concentrations were commonly uninfluenced by whether the tillage was up- and down-slope or cross-slope. Thus, the reduction in soil loss resulting from cross-slope cultivation must be entirely due to the reduction in runoff such cultivation produces.

For plots cultivated across-slope, the step increase in sediment concentration (Figure 6) corresponded to the dominant organization of overland flow into broad, rather shallow channels (typically 20 to 30 mm deep) of preferential flow in a generally down-slope direction. Loch and Donnollan (1983a) refer to these channels as "rills." With plots tilled up- and down-slope, change associated with the step increase in concentration was much less obvious, though undercutting of the furrow sides was noticeable following experiments yielding the higher concentration suspension.

For experiments yielding sediment of higher concentration, apparently additional entrainment mechanisms were brought into play, and the

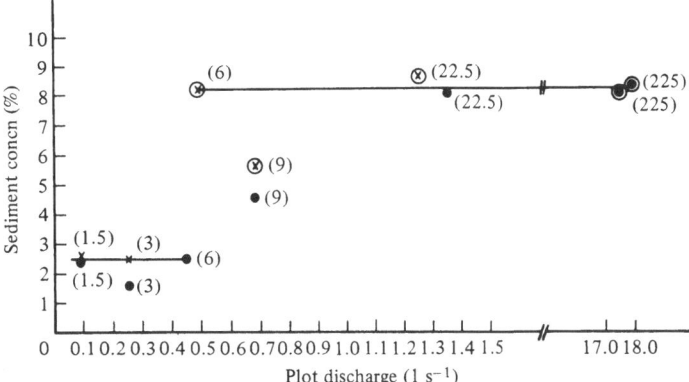

**Figure 6.** Relationship between measured sediment concentration and discharge from field plots of different length subject to constant simulated rainfall. Cross represents cross-slope cultivation; dot, up- and down-slope tillage; circled points indicate rilling observed (Loch and Donnollan, 1983a).

authors describe these as associated with the rilling or channel development observed on cross-slope cultivated plots.

The authors describe the lower of the two concentrations as associated with the rainflow transport mechanism described in the previous section A. However, data given later in this review makes this interpretation uncertain, and it will be argued that runoff entrainment by less channel-organized flow is very likely to be in operation even for the experiments with shorter slope lengths yielding this lower concentration sediment. If this interpretation is correct, then the efficiency of sediment entrainment must be significantly increased (by a factor between 3 and 5 for the two soils investigated) by the processes associated with channelled or rill flow. See section IX.E for an alternative explanation.

This constancy of the sediment concentrations over substantial ranges of plot discharges as illustrated in Figure 6 is a feature noted by other workers under similar experimental conditions.

Despite clay contents of 53% and 73% for the two soils investigated, generally < 10% of sediment was in the clay size range, indicating the aggregated nature of most sediment.

Loch and Donnollan (1983b) found the sediment concentration of different size fractions of sediment to be constant with time. Since selective transport of smaller, more transportable material would be expected to lead to an eventual decline in its concentration, the constancy referred to was taken to indicate constancy of supply resulting from aggregate breakdown under the range of detachment, entrainment, and transport mechanisms operating. Evidence of such aggregate breakdown was obtained.

Loch and Donnollan (1983b) found similar size distributions in the sediment of the two clay soils investigated, yet sediment concentrations were some 1.7 times higher for one soil than another. The authors suggest the higher sediment concentration may be due to lower wet density of aggregates measured for that soil (despite similar ultimate particle densities for both soils).

## C. The Question of Scale

This section has covered research on a scale from a few square meters to a small fraction of a hectare, and the intent is to apply the process concepts to the scale of hectares. It should be noted that such small-scale research is not a scale model of the erosion process but a part of the erosion/deposition system.

Extrapolation of results obtained at small scale to larger scales is not achieved simply by multiplication. Rather, it is achieved through conceptualization of the processes that will operate on the larger target scale and the quantitative representation of these concepts, which we call a model. The next two sections will be concerned with these two steps.

Thus, a role of process models is to act as a synthesizing agent for research at different scales, providing the methodology to allow extrapolation to different scales (Freebairn and Rose, 1982). Valuable though this is, of course, extrapolation always requires experimental testing.

# IV. Conceptualization of Erosion/Deposition Processes

The objective of any model determines the processes of concern and the level of detail at which they are to be represented. The major objective in erosion/deposition modeling is to represent the magnitude and size distribution of sediment flux from an area of land assumed uniform with respect to its inherent and management characteristics that affect this flux. Ways in which processes can be conceived for subsequent quantitative representation to meet this objective will be briefly reviewed. (Selective attention will later be given to situations where some land characteristic, such as slope, varies within the unit of interest.)

## A. Process Conceptualization of Foster et al.

In the model of Foster and Meyer (1972, 1975) and Foster (1982), unit width of downhill flow is hydrologically conceived as containing lateral cross-slope overland flow feeding to a single down-slope channel or rill. The length of the rill segment for the field of uniform characteristics considered here would be the down-slope dimension of the field. The amount of sediment brought to the rill by the cross-slope interrill flow is based on a modified USLE equation. This lateral inflow of sediment is combined with sediment entrained within the rill, to give a "potential sediment load." (In general, this rill segment could receive sediment from an uphill rill segment, but this would not be the case in the uniform field considered here.)

This "potential sediment load" is then compared with a "sediment transport capacity" determined using a modified Yalin bed-load transport equation. Net deposition is assumed to take place in the rill if this transport capacity is exceeded by the potential sediment load. Otherwise sediment transport is limited to the estimated potential load.

The hydrologic concepts used in the model do not apply to the specific lateral/rill flow geometry described above. Even sediment transport, when expressed mathematically (which is how the model is used), simply expresses mass conservation of sediment per unit width of plane: The interrill and rill sediment contributions are simply added without explicit representation of the lateral/rill flow geometry.

Hence, as clarified by Foster (1982) "Flow and sediment are assumed to be uniformly distributed across the slope, and therefore, variables ... are expressed on a total width or area basis although the processes can occur on a limited area."

This clarification does not deny that the lateral/rill flow geometry describes a common field situation. It simply recognizes that without explicit introduction of geometrical features such as number of rills per unit width of plane (which is not attempted), the mathematical model is no different from the assumption of a uniform downslope overland flow, whose depth at any distance down the plane is the spatial average of a very variable quantity.

In application (e.g., in the CREAMS model [Knisell *et al.*, 1980]), the model uses a single characteristic rate of rainfall and runoff for each storm. Foster (1982) describes this as a "quasi-steady" model.

An important implicit assumption in the model of Foster (1982) is that entrainment and deposition process need not be represented separately, but only the *net* of one of these processes over the other will provide an adequate description. Thus, use of terms by Foster (1982) such as "rill erosion rate" $(D_r)$ or (if the same term is negative) "deposition rate" must be interpreted as *net* quantities, the net of erosion over deposition and deposition over erosion, respectively.

Alternatively, the conception may be that no deposition at all takes place, except in situations where "potential sediment load" exceeds the "flow transport capacity." This interpretation follows from Foster (1982, p. 342) who states that in upland erosion sediment particles are not interchanged between the bed and the flow, the reason given being that "the stress required to detach a particle is often considerably greater than that required to transport the particle ... ."

This composition of entrainment (or detachment by rill erosion) and deposition into a single term or the neglect of deposition is perhaps the single most significant conceptual difference between the model of Foster *et al.* and that of Rose *et al.* to be outlined below.

It is because of this composite nature of treating entrainment and deposition that some method has to be introduced to facilitate decision on whether or not net deposition will occur in the down-slope flow. This is achieved by introducing the concept of a "flow transport capacity" ($T_c$) such that net deposition will be taken to occur if this capacity is exceeded by the "potential sediment load" estimated as described above.

Two further assumptions complete the basic concepts in this approach:

1. The net rill erosion (or entrainment) rate ($D_r$) is linearly proportional to the difference between the flow transport capacity ($T_c$) and the total sediment flux ($q_s$) (the same assumption applying to the net deposition rate when this difference is negative).
2. The erosion (or entrainment) capacity ($D_c$) of the rill flow is proportional to the flow transport capacity.

If the constants of proportionality implied in these two assumptions are taken to have the same value, then it immediately follows from these assumptions (Foster, 1982) that

$$D_r/D_c + q_s/T_c = 1$$

or

$$q_s = T_c(1 - D_r/D_c). \qquad [19]$$

Equation [19] implies that in a situation where there is sufficient length of flow for a steady rate situation to be achieved, so that net rate $D_r = 0$, then $q_s = T_c$; otherwise $q_s < T_c$.

The concepts of $T_c$ and $D_c$, while not without meaning, are an outcome rather than a conceptual input in the process conceptualization of Rose et al., now to be outlined.

## B. Process Conceptualization of Rose et al.

As is effectively the case in the model of Foster (1982), so in the model of Rose et al. (1983a, 1983c), no explicit distinction is made between rill and interrill flow. The approximate analytic hydrologic model of Rose et al. (1983b) is used, so the model is dynamic in character, accepting input information in rate form.

Three processes affecting sediment concentration are conceived as acting continuously and therefore simultaneously:

1. Rainfall detachment of sediment, in which raindrops splash sediment from the soil surface into the water of overland flow.
2. Sediment entrainment by overland flow (or runoff detachment); the process whereby flow picks up sediment from the soil surface by the mutual stresses that exist between the soil surface and water flowing over it. The term entrainment is used whether it occurs in rills, between rills, or in sheet flow without rills. Rill or channel flow would commonly be expected to contribute much of the entrained sediment.
3. Sediment deposition, a continually occurring process owing to

sediment settling out under gravity. This process depends sensitively on sediment size or fall velocity.

Processes (1) and (2) increase sediment concentration; process (3) decreases it, as is illustrated in the Forrester-style flowchart of erosion and deposition processes that occur simultaneously at different rates (Figure 7). The resultant sediment concentration ($c$, Figure 7) is determined by the relative magnitude of these different rates, denoted $e$, $r$, and $d$, respectively.

This dynamic systems representation of processes allows the system behavior to emerge automatically from the quantitative representation of processes. Thus, it is quite unnecessary to invoke concepts such as "potential sediment load" or "sediment transport capacity" explicitly, even though system behavior that emerges from the analysis may be described using such terminology, as will be indicated in section VII.

The representation of entrainment and deposition as separate processes marks an important conceptual difference between this model and that of Foster *et al.*, in which only the net result of these two processes is described.

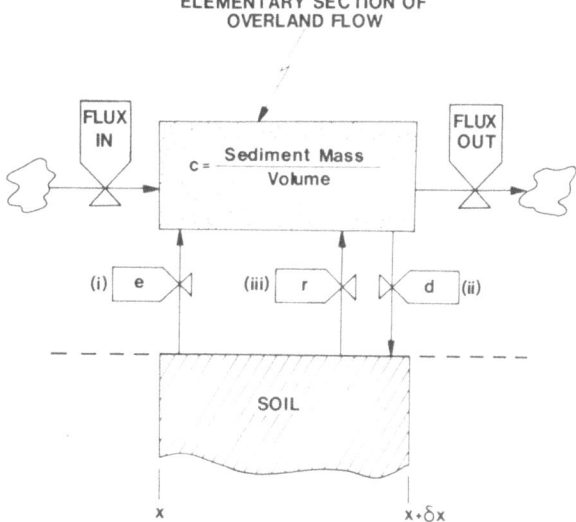

**Figure 7.** Flow chart (after the style of Forrester, 1970) representing the three erosion/deposition processes explicitly represented in the model of Rose *et al.* (1983a, 1983c). Rates shown by valve symbols and direction of flow by arrows. $e$ denotes rate of rainfall detachment, $r_i$ rate of sediment entrainment, and $d$ rate of sediment deposition. Fluxes in and out are sediment fluxes in overland flow, shown artificially elevated above the soil surface to indicate exchange fluxes more clearly Rose *et al.* (1983c). Reproduced from SOIL SCIENCE SOCIETY OF AMERICA JOURNAL, Volume 47, 1983, pages 991–995, by permission of the Soil Science Society of America.

The reason for this separate rather than composite representation is that the two processes depend on different and largely independent factors. For example, rate of entrainment depends on such factors as stream power (i.e., water flux and slope), soil strength characteristics, etc. Rate of deposition, in contrast, is largely uninfluenced by such factors, depending instead dominantly on the size distribution of sedimentary units.

Thus, any correlation found between the *net* of one rate over the other would be expected to be rather specific to the particular experiment and lack the generality desired in process models.

The way in which processes (1) through (3) are quantitatively represented by Rose et al., (1983a, 1983c) is described later. The resulting first-order partial differential equation expressing conservation of mass of sediment is reduced to an ordinary differential equation that can therefore by solved analytically for sediment concentration. Sediment flux follows from equation [1], and this flux is summed over the duration of the erosion event to yield total sediment loss.

At sufficiently low slopes for process (2) (entrainment) to be inoperative, the outcome of processes (1) and (3) is what Moss et al. (1979) described as "rain-flow."

In section VII further comparison between the models of Rose et al. and Foster et al. will be given to that noted here and in subsection A above.

# V. Erosion/Deposition Model of Rose *et al.*

Quantitative expression will first be given to the three erosion/deposition processes, described in section 1VB and illustrated in Figure 7. These expressions will then be combined in an equation that expresses mass conservation of sediment in an overland flow whose hydrologic characteristics are as described in section IIB. This combination provides an analytic model of the erosion/deposition process on a plane land surface.

The analytic solutions obtained are capable of extention to more complex situations, such as a plane of one slope discharging onto a second of different slope. Consideration of greater geometrical complexity will be reserved until section VIII. This question of greater geometrical complexity of surfaces will not be treated exhaustively in this review.

## A. Quantitative Representation of Erosion/Deposition Processes

The same numbering of processes will be used as in their description in subsection IVB, despite a change in order of presentation.

1. Rate of rainfall detachment, $e$, is defined as the mass of soil detached by rainfall per unit area and time. It is described following Rose (1960, 1961) and assuming that $e$ is proportional to the fraction of soil ($C_e$) exposed to raindrops. Thus,

$$e = aC_e P^p,  \quad\quad [20]$$

where $a$ is a measure of the detachability of soil by rainfall of rate $P$; and power $p$ (non-dimensional) will hereafter be approximated by 2. The value of $a$ can be determined (Rose, 1960) and is found to depend on soil type and management and depth of overland flow (Palmer, 1964).

Suppose the sediment is conceptually sorted into a number ($I$) of size ranges, each of equal mass (to simplify calculation). Assuming the rainfall detachment process is unselective with respect to size of sediment detached, then the detachment rate $e_i$ of sediment in the size range $i$ will be given by

$$e_i = e/I. \quad\quad [21]$$

3. Rate of sediment deposition, $d$, is defined as the mass per unit time and bed area of arrival of sediment at the bed owing to the process of sedimentation. This process is here described in different order to its introduction because it illustrates clearly the need to recognize explicitly that sediment is not normally closely sorted and exists in a size range of particles or aggregates of particles.

   The size of sedimentary units has a major effect on rate of sedimentation, so that $d$ must be computed as the sum of $d_i$ calculated separately for each sediment size class $i$. Each size class is defined by its settling velocity $v_i$, assumed measured at an appropriate sediment concentration (Kynch, 1952). Thus,

$$d_i = v_i c_i, \quad\quad [22]$$

where $c_i$ denotes the sediment concentration of size class $i$. The distribution of $v_i$ can be measured for sampled sediment using the "bottom withdrawal tube" method (Anonymous, 1943)

2. Rate of sediment entrainment, $r$ (also described as rate of runoff detachment). Moss *et al.* (1980) found $r$ to be linearly related to stream power ($\Omega$) when sediment transport reached the stage defined as "rheologic flow." This stage would be dominant in an erosion event of field significance. Previously, Bagnold (1977) provided similar evidence that the rate of transport of bed load in streams or rivers could be linearly related to the stream power in excess of a threshold value ($\Omega_0$).

   Bagnold (1977) mathematically described the net rate of bed-load transport in streams, $w$, defined as the immersed weight passing unit width of stream bed per unit time, by

$$w = \eta(\Omega - \Omega_0)/0.6, \quad\quad [23]$$

where $\eta$ (non-dimensional) is the efficiency of bed-load transport, $\Omega_0$ is the value of $\Omega$ at which the entrainment process commences, 0.6 is the friction coefficient for transported sediment (Bagnold, 1977), and

$$\Omega = \rho g S q \qquad [24]$$

with

$$q = Qx,$$

$\rho$ is the density of water, and $g$ acceleration due to gravity.

Converting the immersed weight $w$ to an entrained sediment mass flux $q_s$:

$$\left. \begin{aligned} q_s &= K(\Omega - \Omega_0) \\ &= qc \quad \text{(from equation [1]),} \end{aligned} \right\} \qquad [25]$$

where

$$K = 0.276 \, \eta \qquad [26]$$

and

$$\Omega_0 = \rho g S Q x_*, \qquad [27]$$

$x_*$ being the distance down the plane where entrainment commences (i.e., where $\Omega >$ the threshold value, $\Omega_0$). The number 0.276 in equation [26] assumes a sediment specific gravity of 2.6 and requires modification if this is not so (Rose, 1983c).

It follows from equation [25] with equations [24] and [27] that

$$c = \rho g S K (1 - x_*/x) \qquad [28]$$

and $c_i = c/I$ is assumed.

Equation [28] will be assumed to apply to entrainment in general, not just to bed-load transport as established by Bagnold (1977); hence, $\eta$ should be renamed an efficiency of entrainment.

From Figure 7, mass conservation of sediment in the elementary volume requires that

$$r_i = d_i + \partial q_{si}/\partial x + \partial/\partial t (D c_i),$$

where

$$q_{si} = q c_i,$$

and $D$ = flow depth at any $x$ and $t$.

Substituting into this expression for $r_i$ from equations [22] and [28] (divided by $I$), then

$$r_i = v_i c_i + \rho g S K Q/I + \frac{\partial}{\partial t}(D c_i); \; (x > x_*). \qquad [29]$$

It is assumed that equation [29] will continue to apply, at least approximately, when $c_i$ is modified by the rainfall detachment process. Suppose now that only a fraction $(C_r)$ of the bed is exposed to entrainment by overland flow, the remaining fraction $(1 - C_r)$ being protected by mulch, stone, or other material in effective contact with the soil surface.

In a similar manner to the use of $C_e$ in equation [20] (though without as good justification), let $C_r$ be introduced as a multiplier into equation [28] (so affecting term $v_i c_i$ in equation [9]), and also into the second term on the right-hand side of equation [29]. Then, using equation [28] (thus modified), equation [29] becomes

$$r_i = (\rho g S K Q / I)\left(\gamma_i - \frac{v_i x_*}{Q x}\right)C_r + \frac{\partial}{\partial t}(D c_i), \qquad [30]$$

where $x \geq x_*$ is assumed. In equation [30],

$$\gamma_i = 1 + v_i/Q,$$

and $C_r$ = fraction of soil surface unprotected from entrainment by mulch, stone, or other such material in contact with the soil surface.

Note that the first term on the right-hand side of equation [30] is directly related to stream power, and the second reflects the rate of increase in sediment storage in the elementary section of Figure 7.

Comment should be made on the simple direct factor introduction of $C_r$ (equation [30]). The effect of $C_r$ on $r_i$ is almost certainly more complex than this direct proportionality. It is likely that cover elements in contact with the soil surface will absorb a greater fraction of stream power than the fraction of surface covered by them. An analogy might be isolated trees in a savannah, where the trees would absorb proportionally much more of the shear stress, and so stream power, than the land surface between the trees. The reason for the direct factor introduction of $C_r$ into the theory is to remove the direct fractional area effect, so that the expected greater complexity involved will show up in a non-linear relationship, between $\eta$ and $C_r$, as is illustrated from field experimental data in section IXB.

It follows from equation [30] that $r_i$ depends on $v_i$ through the term $\gamma_i$ equation [31]. In a steady state, and if it is assumed that sufficient flow or value of $x$ exists such that $x_*/x$ is very small, then equation [30] is simplified to

$$r_i = \rho g s K Q C_r \gamma_i / I.$$

Thus, $r_i$ is then proportional to $\gamma_i$ and so to $v_i$ (since $\gamma_i$ is typically $\gg 1$, except for very small sediment; see equation [31]). This proportionality between $r_i$ and $v_i$ is consistent with the assumption that entrainment rate is neutral with respect to the size of bed sediment. This consistency follows since the relative concentration of bed material in size class $i$ will be proportional to the relative deposition rate for sediment of that size (if an agrading bed in dynamic equilibrium is assumed). This question of the relationship between bed and sediment load characteristics will be considered further and the above argument clarified in section VIIIC.

## B. Model of Erosion/Deposition on a Planar Land Element

It will be assumed that the planar land element under consideration has no flux of water entering its up-slope edge, so that $q = 0$ at $x = 0$.

The model follows from consideration of mass conservation of sediment in the elementary section of overland flow (Figure 7) when all processes are acting, combined with marriage of the theories of sediment concentration and hydrology reviewed above.

From Figure 7, mass conservation of sediment of size range class $i$ and concentration $c_i$ requires that

$$\frac{\partial}{\partial x}(qc_i) + \frac{\partial}{\partial t}(Dc_i) = e_i - d_i + r_i,$$  [32]

where the algebraic sum of rates on the right-hand side of equation [32] represents the net erosion rate.

With equation [21] for $e$, equation [22] for $d_i$, and equation [30] for $r_i$, to a good approximation the partial differential equation [32] can be reduced to an ordinary (first-order) differential equation, which is readily solved. This simplification follows from cancellation of the term $\partial/\partial t\,(Dc_i)$ on the left-hand side of equation [32] with the same term on its right-hand side where it is a component of $r_i$ (see equation [29]). This cancellation is approximate, however, since $c_i$ in equation [29] represents sediment concentration in the absence of rainfall detachment, whereas this process is acknowledged through the rate $e_i$ in equation [32]. This approximation is very good if $e_i$ is small compared to $r_i$, and fortunately this is usually the case, as illustrated in section IXA.

Solution of the ordinary differential equation resulting from this simplification of equation [32], with $c_i$ summed over all size classes, yields the sediment concentration $c(L, t)$ at the bottom of the plane of length $L$ as a function of time $t$. The result is

$$c(L,\, t) = (aC_eP^2/QI)\sum_{i=1}^{I}(1/\gamma_i) + \rho gSKC_r(1 - x_*/L) \quad (L > x_*).$$  [33]

The first term on the right-hand side of equation [33] is due to rainfall detachment, and the second term to entrainment, both being net values over deposition.

It follows from equations [33, 1], and $q = Qx$ that the sediment discharge from the bottom of the plane $[q_s(L)]$ is given by

$$q_s(L) = c(L,\, t)QL.$$  [34]

Hence, the accumulated mass of sediment $(M_s)$ lost from a plane of width $W$ is given by

$$M_s = WL \int_0^{t_R} c(L, t)Q(t)dt, \qquad [35]$$

where $t_R$ is the duration of the runoff event.

## C. Discussion of Model of Rose *et al.*

The partial differential equation [32] simplifies to an ordinary differential equation through approximate cancellation of term $\partial/\partial t(Dc_i)$ (Rose et al., 1983c). The resulting ordinary differential equation has coefficients that vary with time. The solution given in equation [33] is to the linear equation obtained assuming the coefficients are constant or have a mean value appropriate to a small time interval (such as 5 minutes) over which the time-varying quantities $P$ and $\gamma_i$ can be so replaced. Thus, in practice, the solution (equation [33]) will be recalculated, giving $c(x, t)$ and $c(L, t)$ in particular (equation [34]), which is summed over time for the whole runoff event to yield total soil loss, as indicated in equation [35].

It may be shown that in major erosion events the second term is considerably greater than the first term on the right-hand side of equation [33]. This leads to a very simple approximation explored in section IX. Only the first term (involving $\gamma_i$) in equation [33] is dependent on the size distribution of sediment, so the importance of this distribution on sediment loss from a plane becomes more significant with reduction in size of the runoff event. From equation [31] and [33] it follows that the concentration of smaller particles, with their correspondingly smaller $v_i$ and so $\gamma_i$, will be enriched in the contribution to sediment concentration made by this term.

The most important unknown in the solution is the entrainment efficiency $\eta$ (in $K = 0.276\eta$). While by definition $0 < \eta < 1$, at present $\eta$ can only be determined accurately by calibrating the model against measurement data on sediment concentration or sediment flux. As will be demonstrated in the next section, such calibration can yield values also for $a$ and $\Omega_0$, though the value of $a$ can be separately determined as mentioned above.

Values of $\eta$ as high as approximately 0.7 have been determined for a bare cultivated vertisol under intense rainfall (Rose *et al.*, 1982a), and as low as approximately 0.05 for an arid zone catchment (Rose *et al.*, 1983d). Values of $\eta$ depend on soil type, management, and management history; values are expected to increase with decreasing shear strength, a substantial increase with cultivation being possible, for example. The increase in sediment concentration associated with the onset of channelled flow observed by Loch and Donnollan (1983a) and illustrated in Figure 6 is interpreted by this model as due to an increase in $\eta$ associated with new entrainment processes introduced by channelled or "rill" flow. The character of entrainment processes, and thus $\eta$, depend on stream power. The dependence of $\eta$ on these and other factors will be illustrated further in section IX. Progress towards an ability to predict $\eta$ is also reviewed there.

Equation [35] indicates that the soil loss per unit land area, $M_s/WL$, does not depend explicitly on the length $(L)$ of the plane. The dependence on slope length is therefore implicit, partly coming through an increase in $\eta$ with slope length resulting from the development of channels or rills and their effectiveness as $L$ increases, thus increasing entrainment efficiency, as mentioned above. Further evidence of dependence of soil loss per unit area on slope length is considered in section IX.

In the tropical environment of West Africa investigated by Lal (1982), soil loss per unit area on slopes of 10 and 15% displayed a broad maximum in the region of 20-m length. Slope lengths commonly used in graded channel or contour bank systems are such that little if any reduction in soil loss per unit area from the cultivated interbank area is achieved, deposition occurring more in the graded channel than in the ultimate drainage system, as quantitatively discussed in section VIII.

Even if contour bank systems achieve no substantial reduction in soil loss per unit area, there are other benefits derived (balanced by costs in construction and maintenance), such as reducing the maximum severity of damage and the likelihood of gully development.

The measurement of $v_i$ necessary to evaluate term $\gamma_i$ in equation [33] can be achieved using a top entry (settling) tube or bottom withdrawal tube (Anonymous, 1943). Since $v_i$ depends on sediment concentration (Kynch, 1952), realistic concentrations should be used on its evaluation. The ability to determine $v_i$ at any concentration may confer some advantages to the bottom withdrawal tube technique over the conventional top entry system, though the two techniques can be used in a complementary way. Taking samples for the determination of $v_i$ soon after a runoff event can reduce uncertainty in such data.

The runoff per unit area $Q$ in equation [33] requires either direct measurement or a knowledge of the site infiltration characteristics, as discussed in section II.

The importance of length $x_*$ is in its relation to $L$ (equation [33]). Since $x_*$ is inversely proportional to $Q$ (equation [27]), it is strongly time variable. As will be illustrated in section IX, the effect of the term $(1 - x_*/L)$ is often small except for slope lengths less than approximately 20 m. Thus, in systems involving mechanized agriculture the term can be ignored without introducing significant error to estimated soil loss except in events where such loss is quite small. The only parameter required to calculate $x_*$ is the threshold stream power $\Omega_0$, and a method of deriving values of $\Omega_0$ from data on soil loss from plots of more than one slope length is given in section IXE. Values of $\Omega_0$ tend to be on the order of 0.01 W m$^{-2}$ for agricultural soils (see section IX and Freebairn and Rose, 1982).

With a stone mulch or surface organic mulch in good contact with the soil surface, the two surface exposure factors $C_r$ and $C_e$ are equal. Quite generally, however,

$$C_e \leq C_r \quad \text{(Rose et al., 1983c).}$$

More detailed consideration of how the data requirements of this model can be obtained is given in Rose et al. (1983a) and Rose (1984).

## VI. Application of Model of Rose *et al.*

Application of the model described in section V is illustrated by data from a small (1.3 ha) watershed or catchment (Watershed 63.101) on the Walnut Gulch Experimental Watershed in Arizona. The watershed can be approximately represented as a plane of length ($L$) 194 m, width ($W$) 67 m, and total relief 7.8 m. Thus, $S$ = 0.0402. Rainfall, runoff, and sediment concentration in runoff were measured as a function of time at the lower boundary of the watershed. Pump-type (suspended sediment) samples were taken every 3 minutes during runoff for the measurement of sediment concentration. $Q$ was calculated from measured water flux $q(L)$ using equation [9].

There was some uncertainty in data on $v_i$, but this uncertainty affected the values of $a$ and $\eta$ derived as described below by a factor less than 1.5 (Rose *et al.*, 1983d).

In this experiment, the value of $\Omega_0$ has negligible influence on total soil loss during erosion events. However, its value can be obtained towards the end of an event, when runoff continues despite cessation of rainfall. Then, with $P = 0$, the first term on the right-hand side of equation [33] is also zero, thus eliminating the possibility of the value chosen for $a$ interacting with that chosen for $\eta$. Also, when $P = 0$, $Q$ is small and thus $x_*$ large compared to $L$ (equation [27]), so that $\Omega_0$ can be determined rather accurately. The value for $\Omega_0$ of 0.00015 W m$^{-2}$ gave good agreement between the theory and data.

In experiments with relatively simple rainfall pattern, measured sediment concentration declined monotonically from initial values of up to about 5 kg m$^{-3}$ measured as early as was feasible with the type of equipment employed. In erosion it is the sediment flux $q_s$ that is of more significance than sediment concentration. Figure 8 illustrates the good agreement obtained for $q_s$ when the flux calculated by the theory is calibrated against the measured flux by varying values of the parameters $a$ and $\eta$.

The values of $a$ and $\eta$ that gave the agreement illustrated in Figure 8 were obtainable with virtually no interaction between them. This independence in evaluation was because the initial high values of sediment concentration were dominated by the value of $a$, because of the high values of $P^2/Q$ pertinent to this initial period (see the first term in equation [33]). During the later period of the event, however, the second term on the right-hand side of equation [33] dominated with $P^2/Q$ much smaller. Unless affected by the term in $x_*$, the second term associated with entrainment makes a constant contribution to $c$ if $\eta$ in term K is constant during the runoff event. In general, $\eta$ is not expected to be strictly constant throughout a runoff

30                                          Calvin W. Rose

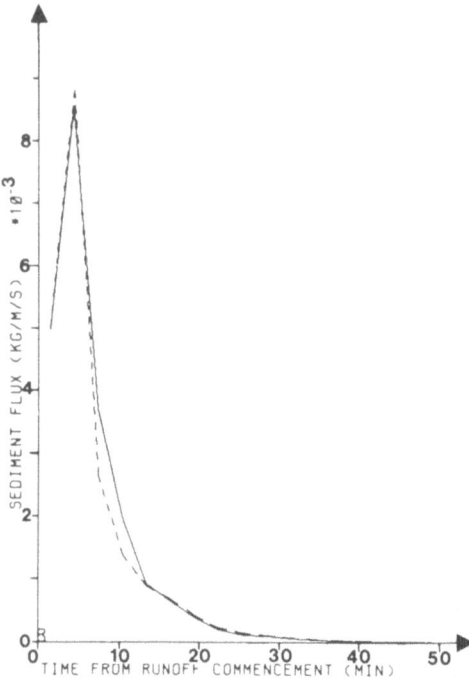

**Figure 8.** Comparison of sediment flux vs. time for experiment 8 described in Rose *et al.* (1983d). Solid line represents measured results; dashed line represents calculated results. Rose *et al.* (1983d). Reproduced from SOIL SCIENCE SOCIETY OF AMERICA JOURNAL, Volume 47, 1983, pages 996–1000, by permission of the Soil Science Society of America.

event (as considered later). However, in these experiments a constant value of $\eta$ gave quite good agreement (Figure 8), with some evidence of a reduction in $\eta$ as $Q$ (and hence stream power) declined towards the later period of the event.

In summary, values of $a$, $\eta$, and $\Omega_0$ could be readily obtained with very little interaction between the values of each parameter that gave good agreement between the theory and data. It would be desirable to determine the value of $a$ quite independently, as discussed in section V, but this independent check is not yet available. A feature of the variation in the value of $a$ required to fit the field data is illustrated in Figure 9. Since the cover factor $C_e$ was not measured, only the product $aC_e$ is available, but this product is still proportional to $a$. Laboratory experimentation (e.g., Palmer, 1964) has indicated that parameter $a$ first increases with depth of water ponded on the soil surface, reaches a maximum at a shallow depth, and then declines at surface water becomes of sufficient depth to provide

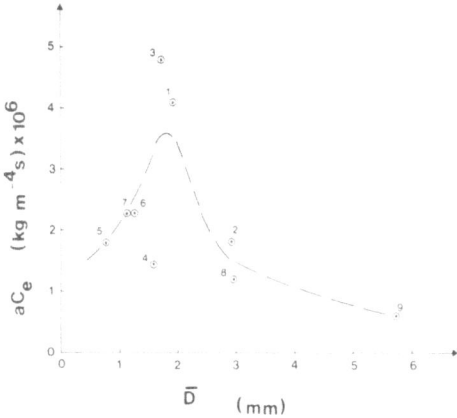

**Figure 9.** The parameter $aC_e$ in the theory of Rose *et al.* (1983c) determined by calibration against data from nine runoff events in an arid zone catchment, shown plotted against $\bar{D}_m$, the mean depth of overland flow for each event. The dashed line hand-fitted to these data indicated the type of variation expected from the laboratory studies of Palmer (1964) Rose *et al.* (1983d). Reproduced from SOIL SCIENCE SOCIETY OF AMERICA JOURNAL, Volume 47, 1983, pages 996–1000, by permission of Soil Science Society of America.

the soil surface with protection from raindrop impact. Moss and Green (1983) noted a similar finding in connection with rain-flow transportation. Figure 9 provides evidence of similar behavior, with intriguing accuracy for field data.

The values of $\eta$ (or, strictly, $\eta C_r$) from this series of nine experiments over a 2-yr period varied by less than a factor 3, and use of the average value of 0.006 would still give good agreement between measured and calculated data. This relative constancy may reflect little change in surface characteristics of the arid zone site over that time period. Indications were that $C_r \cong 0.1$, so the average value of $\eta$ for the experiments would be $\cong 0.06$, typically less than for tilled agriculture soils, as would be expected.

## VII. Comparison of the Modeling Approach of Foster *et al.* and Rose *et al.* for Overland Flow

The approaches adopted by Foster and Rose and their co-workers represent two mainstreams in erosion/deposition modeling. Some of the similarities and differences between these two approaches in process conceptualization for model purposes have already been considered in section IV. The purpose of this section is to carry forward this comparison in more detailed and analytic form.

Comparison of models is never a simple exercise and is fraught with the dangers of ignorance and bias when an author attempts to compare other models with his or her own. However, such a comparison is of potential pratical significance, and the issues are too important to be shirked. I trust ignorance and bias will be corrected in due course.

As explained by Foster and Lane (1981), the model of Foster *et al.* "is intended to be useful without calibration or collection of [further] research data to determine parameter values. Therefore, established relationships, such as the USLE, were modified and used in the model."

Because of reservations in using the USLE in quite different environmental contexts from that in which it was derived, and the difficulties and reservations discussed in section I of other countries repeating this particular research methodology to ensure relevance to their own contexts, this author did not operate within the above restraints accepted by Foster *et al.* The penalty of such freedom is that currently parameter determination is necessary to apply the model of Rose *et al.*, though a framework for parameter prediction is emerging, as discussed later in this review. The opportunity provided by such freedom is to seek theory such that the necessary parameters are of a more fundamental character. Whether this potential advantage is achieved remains for continued evaluation.

In this section comparison is limited to erosion and sediment yield from field-sized areas, where overland flow can usually be well represented by the kinematic flow approximation used in section II. In practice this means that flow depths are sufficiently modest that dynamic terms in the momentum equation are negligible in comparison to other terms.

Specific examples of how these modeling approaches are developed to represent more complex watersheds will receive some attention in later sections.

It is also appreciated that, like all models, that of Foster *et al.* has undergone continuous development and adaption to specific context needs. This comparison has therefore attempted to focus on basic concepts used, rather than the variety of specific forms in which these have been presented (Foster and Meyer, 1975; Foster *et al.*, 1977; Foster and Lane, 1981; Foster, 1982).

## A. Consideration of the Basic Rill Transport Equations of Foster *et al.*

In section IVA it was suggested that the need to introduce the concept of "flow transport capacity" ($Tc$) in the model of Foster *et al.* arose because deposition was not represented as a continuous ongoing process in its own right, as it is in the model of Rose *et al.*

This critique will be amplified by analyzing the situation that illumines the basic equations of Foster *et al.* This situation is depicted in Figure 10 in which sediment-free water flows uniformly from a non-eroding to an eroding surface at $x = 0$. Following Foster and Meyer (1975), Figure 10

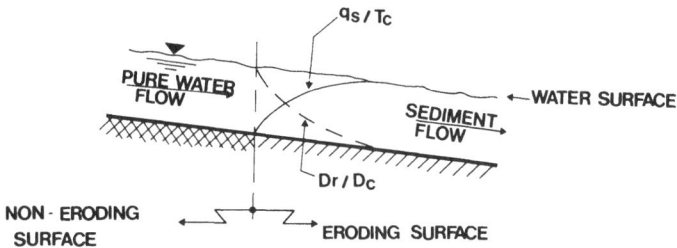

**Figure 10.** Interactions between flow detachment and sediment load as given by the theory of Foster *et al.* See text for symbols. Adapted from Foster and Meyer (1975).

also shows the two ratios $q_s/T_c$ and $D_r/D_c$, where, as shown in equation [19],

$$D_r/D_c + q_s/T_c = 1.$$  [36]

In terms of the symbols used in section V the situation depicted in Figure 10 may be described mathematically by

$$e_i = 0; \quad x = 0; \quad \partial/\partial t \, (Dc_i = 0); \quad Q = 0,$$

and $q$ is constant so that equation [10] does not apply.

The boundary condition is

$$c_i = 0 \quad \text{at } x = 0.$$  [37]

This situation in Figure 10 will now be analyzed using the concepts of the theory of Rose *et al.* (section V), the results of this analysis will then be compared to the basic equation [36] of Foster *et al.*

In flow of depth such that $e = 0$, the remaining processes described in the theory of Rose *et al.* are entrainment and deposition (section IVB). With equations [22] and [30], which describe these processes, equation [32] then becomes

$$\frac{dc_i}{dx} + \frac{v_i}{q} c_i = \frac{\rho g \, SKC_r \, v_i}{Iq}$$  [38]

which is a linear, first-order differential equation, whose solution is

$$c_i = A\left[ 1 - \exp\left( -\frac{v_i}{q} x \right) \right],$$  [39]

where the asymptotic value $A$ in equation [39] is

$$A = \rho g \, SKC_r/I.$$  [40]

Hence, the sediment flux $q_s$ is given by

$$q_s = qc = \rho g q S K C_r \frac{1}{I} \sum_{i=1}^{I} \left[ 1 - \exp\left(\frac{-v_i}{q} x\right) \right]. \qquad [41]$$

Comparing equations [41] and [19] it can be seen that the "flow transport capacity" $T_c$, introduced as a concept to be separately evaluated, is nothing more than the asymptotic solution for $q_s$ (equation [41]), so that

$$T_c = \rho g q S K C_r. \qquad [42]$$

Thus, introduction of the term $T_c$ into the analysis is consistent with the physical expectation in relation to Figure 10, expressed also in the outcome of the above analysis, that ultimately an equilibrium sediment flux will be achieved if factors affecting entrainment and deposition remain unaltered.

Furthermore, without the need to introduce the concept of "rill erosion detachment capacity rate" $(D_c)$, or to make the assumption (Foster and Meyer, 1975) that "rill erosion rate" $(D_r)$ is proportional to the difference $(T_c - q_s)$, an expression for the variation with $x$ of $D_r/D_c$ follows from the comparison of equations [19] and [41].

The above analysis and discussion are the justification for the statement at the end of section IVA that using theory of the type illustrated by Rose *et al.*, the concepts $T_c$ and $D_c$ need not be introduced as conceptual inputs. Rather, $T_c$ and $D_r/D_c$ emerge from the analysis, together with analytic defining expressions for them.

From this analysis (equation [41]), the physical meaning ascribable to $T_c$ is the equilibrium sediment flux approached asymptotically in the manner described by equation [41]. Also from equations [19] and [41], at any $x$, $(T_c D_r/D_c)$ is the deficit in $q_s$ below the equilibrium sediment flux $(T_c)$.

Substitution of expressions of the forms derived by Foster *et al.* for the terms in equation [19] reveal some discrepancies and similarities to equation [41] for $q_s$. However, perhaps the most practically important difference between the two approaches is the lack of necessity to introduce *a priori* concepts, such as $T_c$ and $D_c$, into the analysis if deposition and entrainment are conceived of as described in section IVB. Thus, it appears that the modeling approach of Rose *et al.* is conceptually simpler in terms of required concepts, even if the end expressions reached are not startlingly different.

## B. Comparison of Specific Forms of Foster *et al.* with Rose *et al.*

Comparison of the two models here is sought using as far as possible the "rill" and "interrill" concept of Foster *et al.* In broad terms, it is possible to view the first term on the right-hand side of equation [33] as the component contribution to sediment concentration arising from interrill processes, in that it is the net contribution of rainfall detachment over deposition (as defined in section IVB). However, this term corresponds more closely to the "rain-flow" concept of Moss *et al.* (1979) than to the contribution of all interrill processes, since in general even interrill flow can exceed the threshold stream power $(\Omega_0)$, which may be lowered by the effect of raindrop impact in the shallower flow typical of the interrill region.

The second term in equation [33] may be more closely identified with rill processes in those (common) situations where some form of rilling exists. However, for reasons given in the above paragraph, this term may also include some contribution arising from interrill flow.

Foster (1982) has recognized the difficulty in clearly separating interrill and rill effects. Using the terminology of Rose *et al.* (of section IVB), recognition that in addition to rainfall detachment and deposition, entrainment can also take place in interrill flow appears to have led to the two basic forms used by Foster *et al.* to represent interrill erosion rate $(D_I)$. Sometimes $D_I$ is represented as proportional to rainfall rate $P$, or $P^2$. Perhaps more commonly, however, $D_I$ is represented with additional terms that come from the USLE, which include a dependence on slope $(S)$ and a measure of storm rainfall energy $(EI_{30})$. Since the original USLE aimed to represent sediment loss by all processes, including rills, the USLE is modified for use in this context (Foster, 1982), presumably by removal of the "rill effect" from it. Foster (1982) discusses these difficulties.

Foster (1982) defined interrill $(D_I)$ and rill $(D_r)$ erosion rates as

$$\frac{dq_s}{dx} = D_I + D_r. \qquad [43]$$

Hence, from [1]

$$\frac{dq_s}{dx} = \frac{qdc}{dx} + \frac{cdq}{dx}$$

$$[44]$$

$$= \frac{qdc}{dx} + cQ,$$

from $q = Qx$.

If $c$ is conceived as having an interrill $(c_I)$ and rill $(c_R)$ component, and $c_I$ is assumed to be independent of $x$, then equation [44] may be written in the following form to aid comparison with equation [43]:

$$\frac{dq_s}{dx} \doteq c_I Q + \frac{qdc_R}{dx} + c_R Q. \qquad [45]$$

In the steady-state situation here assumed, the theory of Rose *et al.* would be

$$\frac{dq_s}{dx} = e + r - d. \qquad [46]$$

It is not straightforward to compare equations [45] and [46]. However, Rose et al. (1983c) have attempted such a comparison for a particular form of the model of Foster et al., essentially assuming $e = Qc_l$ and that $(r-d)$ is given by the remaining terms. Some similarities were noted, but there were significant differences, both in slope dependence and arising from the representation of deposition as an ongoing process with $d = \Sigma_{i=1}^{i} v_i c_i$ (equation [22]).

Thus, it appears that in comparing the two modeling approaches of Foster et al. and Rose et al. differences in concept will have to be evaluated as well as comparison in quantitative terms. Quantitative comparison is made difficult, though not impossible, by the current desirability of deriving at least one parameter ($\eta$ in the case of Rose et al.) from data, and the consequent good fit between the model and data. However, experience is gradually being gained on the physical reality and meaning of such parameters. Some experience with respect to $\eta$ is given in section IX of this review. In due course, such study should lead to an evaluation of the utility of alternative models, both in terms of how basic and simple the concepts required may be and in terms of the adequacy of predictive accuracy.

## C. On Sediment Transport Formulae

Foster et al. uses a modified version of the Yalin sediment transport equation (or series of equations) to estimate $T_c$. This complex equation does contain a dependence on $(\tau - \tau_{cr})$, the difference between the bed shear stress and a critical value of this stress. As Foster (1982) notes, given the uncertainties in estimating flow and sediment characteristics, the distinction between bed-load and suspended load transport, made in some equations, is an unwarranted refinement in shallow overland flow. Use of the Yalin-type equation in such situations was justified by comparing a number of competing equations with a range of flume and field data (Alonso et al., 1981).

The product $qc$, where $c$ is given by equation [33], is the sediment transport model for overland flow proposed by Rose et al. (1983c). However, as section VA indicates, the expression developed for rate of sediment entrainment is closely dependent on the stream power model of bed-load transport by Bagnold (1977). This model was not one tested by Alonso et al. (1981), possibly because as yet no general predictive model to determine $\eta$ is available. It is clear from the data reviewed by Alonso et al. that $\eta$ increases with stream power. Further consideration of factors affecting $\eta$ will be considered in section IX.

Regarding transport rate of sediment as dependent on the rate of work done against the bed shear stresses, general physical principles would seem to favor the dynamic concept of stream power (the rate of working of the bed shear stress) over the static concept of excess shear stress $(\tau - \tau_{cr})$. Since $\Omega = \tau V$, where $V$ is stream velocity, the correlation between $\Omega$ [or $(\Omega$

$- \Omega_0$)], and $(\tau - \tau_{cr})$ makes establishment of one approach as superior to the other more difficult (but see Figure 5).

## VIII. Sediment Deposition and Load-Bed Relationships

Section VB gave the theory of erosion/deposition of Rose *et al.* (1983c) for a planar land element (equation [33]). The first term on the right-hand side of this equation is likely to be of more significance in shallow interrill flow. The contribution of this first term to sediment concentration is dependent on sediment size, being proportional to $1/(1+v_i/Q)$ for any size range i. Thus, particles of small $v_i$ (and so small size) will be better represented. This preferential small-size effect (or else preferential aggregate breakdown) is evident in samples from sediment on short plots (Loch and Donnollan, 1983b, Tables 1 and 2). Sediment in these short plots without rills had approximately twice the clay and silt percentage of sediment sampled from longer plots with rills for both soil types investigated.

To represent sediment transport in a field, including a field with soil conservation structures, the effect of greater geometrical complexity than a plane requires analysis. In subsections A and B below, the same erosion and deposition concepts as those given in sections IVB, and hence the same analytical solution as given in section V, will be applied to a limited range of field features where greater geometrical complexity is accompanied by net deposition of sediment lost from a planar land element.

This generality of application of the solution given in equation [33] demonstrates the power of the basic definition of processes involved combined with the use of analytic rather than numerical solutions (since this is possible here).

In subsection C relationships between the size distribution of sediment in the moving load compared to that of the bed are considered.

### A. Sediment Deposition in a Channel or Stream Receiving Sediment along Its Length

The geometry of the situation considered here is illustrated in Figure 11. The channel may be formed by a graded channel terrace (or contour bank) with slope substantially less than the plane from which sediment is received. (A very similar solution applies if the channel were a shallow stream, though upstream boundary conditions are different in this case.)

Let $q_p, q_c$ = runoff rate per unit width of plane and channel respectively $(m^3 m^{-1} s^{-1}$; i.e., $m^2 s^{-1})$,

$F_c$ = volumetric flow rate in the channel at exit (Figure 11), $(m^3 s^{-1})$,

$Q_c$ = runoff rate per unit area of channel $(m s^{-1})$, and

$C$ = average width of channel or stream $(m)$.

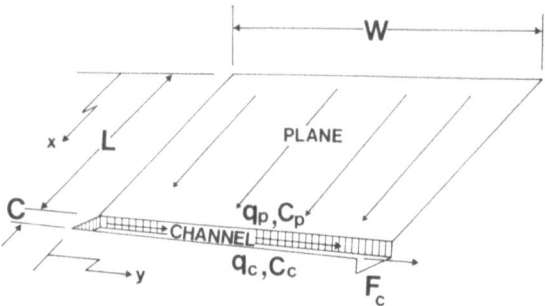

**Figure 11.** The flow of water and sediment from a plane land element (suffix *p*) to a channel (suffix *c*) as in a graded channel system.

Note that the suffix *p* refers to the plane and *c* to the channel. The channel is geometrically approximated as a (long narrow) plane of slope $S_c$, where $S_c < S_p$, the slope of the plane. With trivial generalization of geometrical terms referring to the plane, the analysis below also holds if the surface supplying sediment to the channel is not planar.

The same processes of deposition and entrainment occur in the channel as on the plane. The only differences between sediment flux in the channel and the plane are:

1. Channel flow is assumed to be sufficiently deep that rainfall detachment can be neglected (i.e., $e = 0$). On the other hand, flow depth is assumed not so great that the hydrologic theory of section II would be inadequte.
2. The channel receives a lateral sediment flux from the plane. Since the lateral water flux is large compared with rainfall received directly by the channel, for simplicity the rainfall and infiltration rates for the channel will be assumed equal so that the "excess rainfall rate" is entirely due to water from the plane.

Thus, the model of sediment transport and exchange given in Figure 7 for a plane is also applicable to the channel of Figure 11 provided the rate process shown as *e* in Figure 7 is now considered to be *p*, the rate of sediment transport from the plane to unit area of channel, where

$$p = q_p c_p / C, \qquad [47]$$

with $c_p$ = sediment concentration at exit from the plane.

In parallel, term $Q$ will need to be interpreted as $Q_c$, $x$ as $y$, and $x_*$ as $y_*$, the threshold distance down the channel for entrainment to commence.

The solution given in equation [33] summed over all size ranges *i* is as follows for a particular *i*:

$$c_{pi}(L, t) = aC_e P^2 / Q\, \gamma_i I + (\rho g S K C_r / I)(1 - x_* / L), \; L > x_*. \quad [48]$$

The corresponding solution for the channel sediment concentration, using [47], will therefore be

$$c_{ci}(W, t) = q_p c_{pi}(L, t)/CQ_c \gamma_i I + (\rho g S_c K_c C_{rc}/I)(1 - y_*/W),$$
$$W > y_*.  \quad [49]$$

The total sediment of size range $i$ lost at exit from the channel in a runoff event of duration $t_r$ will therefore be

$$\int_0^{t_R} F_c c_{ci}(W, t) \, dt = CW \int_0^{t_R} Q_c c_{ci}(W, t) \, dt$$

$$= CW \int_0^{t_R} [q_p c_{pi}(L, t)/C\gamma_i I$$
$$+ (\rho g S_c K_c C_{rc} Q_c/I)(1 - y_*/W)] \, dt. \quad [50]$$

The total sediment input to the channel from the plane (of size range $i$) will be

$$W \int_0^{t_R} q_p c_{pi}(L, t) \, dt. \quad [51]$$

$$\text{Let } \phi = \frac{\text{Sediment lost from channel}}{\text{Sediment input to channel}}.$$

Then, dividing equation [50] by [51], and deleting the integrals since both are from 0 to $t_R$, gives $\phi_i$, the value of $\phi$ for sediment size class $i$, as

$$\phi_i = C \left[ 1/C \gamma_i I + \left( \frac{\rho g S_c K_c C_{rc} Q_c/I}{q_p c_{pi}(L, t)} \right) \left( 1 - y_*/W \right) \right]. \quad [52]$$

Now using the approximation that on the plane the first term in equation [33] is negligible compared with the second, then

$$c_{pi}(L, t) = \rho g S_p K_p C_{rp} (1 - x_*/L), \quad [53]$$

where
$$K_p = 0.276 \, \eta_p. \quad [54]$$

Also note that $\gamma_i$ for the channel is given by

$$\gamma_i = 1 + v_i/Q_c. \quad [55]$$

Substituting from equations [53], [54], and [55] into [52] gives

$$\phi_i = \frac{Qc}{I(Q_c + v_i)} + \frac{S_c \eta_c C_{rc} Q_c C (1 - y_*/W)}{S_p \eta_p C_{rp} q_p (1 - x_*/L)}. \quad [56]$$

While equation [56] is general, it may be shown that for graded contour banks of normal design, slope $S_c$ is so low that $y_* > W$, and so the second term on the right-hand side of equation [56] is zero. (Specific figures to illustrate the validity of this statement are given below.) Thus, in this context we obtain the simple relationship:

$$\phi = (1/I) \sum_{i=1}^{I} [Q_c/(Q_c + v_i)].$$  [57]

Foster *et al.* in Chapter 3 of Knisel (1980) give an alternative analysis for the situation considered here.

### 1. Illustrative Example

To illustrate the above theory, data are taken from Loch and Donnollon (1983b) for experiments on a vertisol (Irving clay, a Udic Pellustert). Table 2 from that reference gives data on sediment size distribution, and the mean of figures quoted for plots with rills is used here to yield data on $v_i$ in equation [57].

In erosion events, an average of 14% of the sediment shed by the plane was lost by discharge from the channel at 0.3% slope (Loch and Donnollan, 1983b). In such events, $Q_c$ is likely to be in the range 25 to 50 mm hr$^{-1}$. Using the data on $v_i$ referred to above, equation [57] yields $\phi$ values, expressed as a percentage, of 13.0 and 15.2% for $Q_c = 25$ and 50 mm hr$^{-1}$, respectively. These values enclose the mean measured value of approximately 14%.

To illustrate justification of neglect in this application of the second term on the right-hand side of equation [56], the values of channel stream power $\Omega_c$ at $(2-4) \times 10^{-4}$W m$^{-2}$ are much smaller than the threshold value $\Omega_0$ of $10^{-2}$W m$^{-2}$ determined for this soil by Freebairn and Rose (1982).

### 2. Comment on Above Analysis

Neglecting the second term on the right-hand side of equation [49], as is appropriate for most graded channel terraces, it can be seen that sediment concentration in the channel is not dependant on $y$, the distance down the channel. Hence, rate of sediment deposition in the channel, $v_i c_{ci}$ (equation [22]), will be uniform along the channel. This is consistent with observations that, apart from possible local flow-modified fan-type accumulations at the foot of major rills, there is no evidence of a dependence of depth of accumulated sediment with distance down the channel.[1]

From equation [49] (neglecting the second term) and equation [55], it follows that

$$c_c = \frac{1}{IC} \sum_{i=1}^{I} \frac{q_p c_{pi}}{Q_c + v_i}.$$  [58]

From equation [57] it may be noted that $\phi_i$ would tend to unity (i.e.,

---

[1]Personal communication, D.M. Freebairn.

complete loss of sediment from the channel) as $v_i \rightarrow 0$. At the other extreme, $\phi_i \rightarrow 0$ if $v_i \gg Q_c$, indicating negligible loss of large aggregates from the channel.

## B. Sediment Deposition in a Conveyancing Channel with No Lateral Sediment Input

In fields with soil conservation structures, it is quite common for sediment, for example from a graded channel considered in $A$, to flow through a channel with negligible lateral input of sediment. It is assumed that $e = r = 0$.

Considering mass conservation of sediment through an elementary length of such a channel with steady flow $q_c$ in the $y$ direction leads directly to the simple equation

$$\frac{dc_i}{dy} = -\frac{v_i}{q_c} c_i, \qquad [59]$$

where $q_c = F_c/C$, with $F_c$ the (constant) water flux through the channel.

Integration of equation [59] from entry to the channel section, where the concentration is $c_{1i}$, to outlet distance $Y$ downstream, where the concentration is $c_{2i}$, yields

$$c_{2i} = c_{1i} \exp(-v_i CY/F_c). \qquad [60]$$

Equation [60] represents a simple exponential decline at a rate that depends, among other factors, on $v_i$. Total concentration $c_2$ follows from summation over $i$.

## C. Sediment Size Relationship Between Bed and Moving Sediment Load

Meynink (1983) has developed an equation that, at the stage of dynamic equilibrium in well-developed sediment transport, gives the relative size characteristics of the "bed" and "load," where the "load" is any sediment that is transported. The "bed" refers to the layer within the bed whose size characteristics are affected by the size characteristics of the load.

This relationship was developed to predict the longitudinal deposition patterns of mine tailings inputs to rivers and has recieved experimental confirmation in that context.

However, a quite different and simpler derivation of this equation from that of Meynink (1983) is given below, which indicates that it is likely to be of much more general application than to agrading rivers. Currently, however, this is the extent of its experimental testing.

Let suffix $i$ refer to the sediment size class and suffix $b$ refer to the bed. With reference to the *load*:

Let $c_i$ = load sediment concentration in size class $i$ (average over load),
$c$ = total (average) load concentration

$$\equiv \sum c_i, \text{ where the summation is over } i,$$

$f_i = c_i/c,$

$v_i =$ fall velocity of sediment contributing to $c_i$,

$\bar{v} =$ a mass–weighted average fall velocity defined by

$$\bar{v} = \sum \left( \frac{c_i}{c} \right) v_i, \qquad\qquad [61]$$

$d_i =$ rate of deposition of load sediment in class $i$ expressed as flux density to the bed; then

$$= v_i c_i. \qquad\qquad [22]$$

Also let $d \equiv \sum d_i$, the summation being over $i$,

$r_i =$ rate of entrainment of sediment in class $i$ from the bed, expressed as flux density of sediment leaving the bed due to this process, and

$r \equiv \sum r_i$, the summation being over $i$.

With reference to the *bed*:

Let $c_{bi} =$ bed sediment concentration in size class $i$,

$c_b =$ total bed concentration

$$\equiv \sum c_{bi}, \text{ and}$$

$f_{bi} = c_{bi}/c_b.$

In terms of these symbols, Meynink's equation is

$$f_{bi} = \frac{v_i}{\bar{v}} f_i. \qquad\qquad [62]$$

The derivation of this equation will be shown to follow from equations [61] and [22], and the following two assumptions:

1. A steady rate or dynamic equilibrium exists in exchanges between the load and bed sediments.
2. Entrainment rate is neutral with respect to the size of bed sediment.

Assumption 2 can be expressed by the equation

$$r_i = k c_{bi}, \qquad\qquad [63]$$

where $k$ is a constant not depending on $i$.

Summing equation [63] over $i$:

$$r = k c_b, \qquad\qquad [64]$$

whence, from equations [63] and [64],

$$\frac{r_i}{r} = \frac{c_{bi}}{c_b}. \qquad\qquad [65]$$

Assumption 1 implies no change with time in the size characteristics of the bed or load, which requires that

$$\frac{r_i}{r} = \frac{d_i}{d}.$$ [66]

It follows from equations [65] and [66] that

$$\frac{c_{bi}}{c_b} = \frac{d_i}{d}.$$ [67]

From equations [61] and [22],

$$\bar{v} = (1/c) \sum v_i c_i = d/c.$$ [68]

Substituting from equations [22] and [68] for $d_i$ and $d$, respectively, into [67] yields

$$\frac{c_{bi}}{c_b} = \frac{v_i c_i}{\bar{v} c},$$ [69]

which is Meynink's equation [62].

Assumption 1 would seem to have some generality in that there is experimental evidence that adjustment to a departure from dynamic equilibrium is a very rapid process. Assumption 2 has some conceptual justification. It is difficult to see how the Reynolds or eddy shearing stresses characteristic of more turbulent flow could be selective with respect to sediment size in the entrainment process. However, saltation has been identified as a sediment transport process of significance at lower stream powers (section III), and in wind erosion at least, size selectivity is well known. Thus, while assumption 2 may be more uncertain than 1, equation [69] is a direct consequence of these assumptions.

From the form of equation [69] it follows that a knowledge of size characteristics of the load allows prediction of the relative bed size–concentration properties. However, a knowledge of bed properties does not allow a prediction of load size properties because these include two ratios, for velocity and concentration, not just one as for the bed.

Equation [69] explains the common observation that the bed of streams consists of coarser material than that of the sediment in the load in equilibrium with it. No doubt this occurs also in overland flow, leading to a coarser bed layer.

The Meynink equation would be more likely to hold in the situations of "bed-load amelioration" in overland flow described in section III than in a situation of extreme rilling involving constant cutting into new bed material.

## IX. Simplified Theory of Rose *et al.* with Applications

This section presents a simplified form of equation [33] that provides a good approximation under many situations. It also considers application of this simplified theory to data from a variety of field and controlled plot experiments. These applications illustrate the dependence of the efficiency of sediment entrainment ($\eta$) on a variety of management, soil, and flow factors. The final subsection attempts to synthesize this experience on the variation in $\eta$ to provide the framework for a predictive model for this factor.

### A. The Simplified Theory of Rose *et al.*

The larger the runoff event and the larger the sediment particles or aggregates, the smaller is the first term on the right-hand side of equation [33] compared with the second term. This is because the larger $Q$ and $v_i$ (in $\gamma_i = 1 + v_i/Q$), the smaller this first term becomes (equation [33]). It will be illustrated below that for runoff events associated with "significant" erosion losses, and for well-structured soils, the contribution to sediment concentration made by this term, which represents the net result of rainfall detachment over deposition, can be neglected in comparison with the second term in equation [33]. At least for shallow flows, rainfall rate can indirectly affect $c$ through its effect on $\eta$ (investigated in subsection C below).

This relative magnitude of the two terms in equation [33] may be illustrated using the data of Freebairn and Rose (1982) for $\eta$ and that of Loch and Donnollan (1983b) for sediment size (and hence $v_i$). Assume a major runoff event, with typical values of $P = 50$ mm hr$^{-1}$ and $Q = 30$ mm hr$^{-1}$, say. Then for a slope of 6% typical for the region, the second term in equation [33] is some $10^3$ times greater than the first term. Admittedly, this is a rather extreme example, and the ratio of the two terms changes quickly with hydraulic and soil conditions. However, the example does justify the neglect of the first term in significant erosion events, and in fact even for quite modest erosion events neglect of the first term in equation [33] still provides an approximation whose accuracy is commonly better than the inherent uncertainty in data used to test it.

Thus, the simplified theory that is the topic of this section is given by

$$c(L, t) = \rho g \, SKC_r(1 - x_*)$$
$$= 2700 \, S\eta \, C_r(1 - x_*/L)(\text{kg m}^{-3}), \qquad [70]$$

since $K = 0.276\eta$.

In equation [70] $c$ is a function of time, $t$, because $x_*$ is time variable (equation [27]). However, if $x_*$ is replaced by a time-averaged value, $\bar{x}_*$ (defined in equation [76] below), and if $\eta$ is assumed constant then $c$ is a function only of $L$:

$$c(L) = 2700 \ S\eta \ C_r(1 - \bar{x}_*/L)(\text{kg m}^{-3}). \qquad [71]$$

Furthermore, it may be shown (and will be illustrated in subsection E below) that, provided slope length $L$ is not less than about 30 m, then for most of the time during a substantial erosion event, $x_* << L$. At early and late times during a runoff event, when overland flow is small and $x_*$ not negligible compared with $L$, the sediment loss is only a small fraction of that lost in the entire event. Thus, with the above approximate restriction, $c$ is

$$c = \rho g \ SKC_r \qquad\qquad (L \nleqslant \sim 30 \text{ m}),$$

$$= 2700 \ S\eta \ C_r \qquad\qquad (\text{kg m}^{-3}), \qquad [72]$$

With the assumptions implied in equation [72], sediment concentration $c$ is proportional to $\eta$. If $\eta$ is assumed to have a constant average value during a runoff event, even though under natural rainfall it is likely to be somewhat time variable for reasons discussed in later subsections, then $c$ is constant for any given $S$ and $C_r$ (equation [72]).

There is considerable experimental support for an approximately constant sediment concentration with time. This was illustrated in Figure 6 for data from Loch and Donnollan (1983a), who used a rainfall simulator on field soil. Kilinc and Richardson (1973) also showed that under constant simulated rainfall sediment concentration did not vary significantly with time. Hence these simplified equations have experimental support under constant-rate rainfall. Possible effects of varying rates of rainfall and runoff will be investigated in later subsections.

With $c$ independent of time, equation [35] can be simplified to

$$M_s/WL = c \int_0^{t_R} Q dt, \qquad [73]$$

where $\int_0^{t_R} Q dt$ is total runoff per unit area $(m)$ for the event.

From equations [72] and [73], the time average value of $\eta$ can be calculated from

$$\eta = \frac{(M_s/WL)}{2700 \ SC_r \int_0^{t_R} Q dt} \qquad [74]$$

where $(M_s/WL)$ is the soil loss per unit area (kg m$^{-2}$).

If $(M_s/WL)$ is in Mg (or tonne) ha$^{-1}$, and $\int_0^{t_R} Q dt$ in mm, then the dimensional number in the denominator of equation [74] is 27 instead of 2700.

The simplified equations [80], [72], and [74], will be applied to a range of data in the following sections with the particular objective of elucidating the factors on which $\eta$ depends.

## B. Effects of Cover and Tillage on $\eta$

In field experiments in Queensland described by Freebairn and Boughton (1981) and Freebairn and Rose (1982), surface runoff and sediment loss were measured for a variety of soil surface cover conditions for two vertisols (a black earth or Pellustert and gray clay or Chromustert).

Treatments included a bare fallow, stubble incorporated by disc cultivation, stubble mulch with sweep tine cultivation designed to maintain maximum surface crop residue, and zero tillage. The exposure fraction $C_r$ (equation [71]) was estimated for each treatment. Slopes varied from 5 to 7% (i.e., $S = 0.05$ to $0.07$), slope lengths varied from 35 to 50 m, and measured extreme events included soil loss on the order of 100 Mg (or tonne) ha$^{-1}$ in a single event for bare soil. Sediment loss was the sum of that measured with runoff at exit from the graded channel, as in Figure 11, and the manually estimated sediment deposit in the channel. Measurement of water and sediment loss allowed calculation of $c$, so that $\eta$ was available as the only unknown in equation [72].

The range of surface management treatments described above included both types of tillage and a wide range of surface cover $(1 - C_r)$. $C_r$ was the dominant variable on which calculated $\eta$ depended, and the relationship for the two soil types for the events in 6 yrs of experiments is shown in Figure 12. Scatter in results may be from soil type or cultivation effects, difficulty in estimating soil loss, and other factors.

Likely reasons for the strong non-linearity in this relationship were discussed in section VA. The value of $\eta$ fell from $= 0.7$ at $C_r = 1$ to $\simeq 0.25$ at $C_r = 0.9$. This indicates that only a 10% cover reduced soil loss almost threefold, indicating the well-known conservation effectiveness of stubble cover.

The relationships in Figure 12 [interpreted in terms of cover fraction $(1 - C_r)$] are adequately fitted by

$$\eta = 0.7 \exp\left[-15\left(1 - C_r\right)\right]. \qquad [75a]$$

If equation [75a] is generalized to

$$\eta = b_1 \exp\left[-b_2\left(1 - C_r\right)\right], \qquad [75b]$$

then $b_2$ is likely to be much less site specific than $b_1$.

As indicated by this example, the theory of equations [72] and [74] is so simple that an event-averaged value of $\eta$ can be calculated from any data reporting total runoff, soil loss per unit area, and the fraction $(C_r)$ of soil surface exposed to overland flow. Since the great significance of $C_r$ (as distinct, in general, from $C_e$) has not been made so explicit prior to the development of this theory, there is a great deal of published data in which $C_r$ is not explicitly reported, though the crop grown or surface management treatment is reported. When data are also recorded for water and soil loss from bare soil, for which $C_r = 1$ by definition, then the effective value of $C_r$ can be usefully inferred for each treatment investigated.

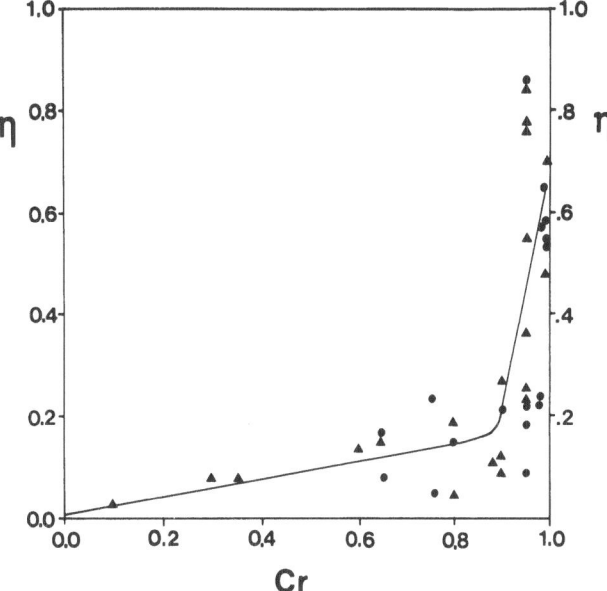

**Figure 12.** Efficiency of entrainment (η) vs. soil surface exposure fraction for two vertisols in the Darling Downs, Queensland. ▲ represents a Pellustert; •, a Chromustert (Freebairn and Rose, 1982).

To illustrate this, Gumbs and Lindsay (1982) measured runoff and soil loss for a major soil type (Orthoxic Tropudult) in Trinidad, with tillage and no-tillage, for bare soil, maize (*Zea mays*), and cowpea (*Vigna sinensis*), over a range of slopes and for two seasonal periods. Plot length was only 7 m. Using equation [74], the average value of η for all bare soil treatments was 0.063, whereas the average value of η $C_r$ for all cropped treatments was 0.011. Taking η = 0.063, the bare soil value, then $C_r$ = 0.18, or the average percentage cover provided by the crops was 82%. The data showed cowpea as providing more effective cover than maize, and for bare soil the value of η for no-tillage was only 61% of the value with tillage.

## C. Effect of Rainfall Rate on η with Shallow Overland Flows

With relatively shallow flows (of calculated mean depth a few millimeters or less), it is commonly observed that sediment concentration and thus soil loss is substantially dependent on rainfall rate (e.g., Singer and Walker, 1983; Kinnell, 1982).

A classic laboratory study of shallow flows under simulated rainfall was carried out by Kilinc and Richardson (1973). One of their conclusions of particular interest to this review was that stream power gave better prediction of sediment discharge than boundary shear stress between the flow and the bed. However, their data are used here to illustrate the dependence of $\eta$ on rainfall rate alone for a small and constant depth of overland flow. Varying rainfall rate alone also varies depth of flow. However, in these experiments (on a bare soil containing 90% sand) slope was also varied. Values of $\eta$ calculated from equation [72] were plotted against mean flow depth for different rainfall rates and slopes. By fitting curves to this data, the value of $\eta$ at a constant flow depth for different rainfall rates could be read off. The results in Figure 13 show a quite linear dependence of $\eta$ on rainfall rate, at least above a rainfall rate of about 25 mm hr$^{-1}$.

It should be noted that Figure 13 is for a mean flow depth ($\bar{D}$) of only 0.53 mm. All experimental results were for $\bar{D}$ less than 0.9 mm. Over the approximate range of 0.4 to 0.9 mm for $\bar{D}$, for any rainfall rate $\eta$ decreased with increase in $\bar{D}$. There was some evidence of a maximum in $\eta$ for $\bar{D}$ between 0.3 and 0.4 mm.

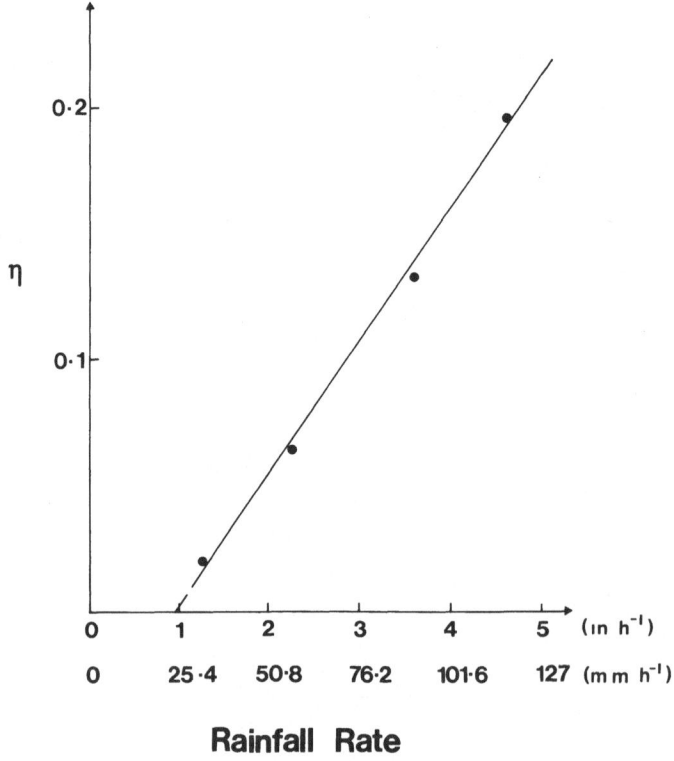

**Rainfall Rate**

**Figure 13.** Entrainment efficiency $\eta$ calculated from the laboratory data of Kilinc and Richardson (1973), shown for four experiments under simulated rainfall of different rate. Flow depth was shallow (0.53 mm) and the same for each experiment.

The rate of increase in η with rainfall rate over the range investigated was not strongly dependent on $\bar{D}$ over the experimental range. However, the "threshold" rainfall rate resulting from extrapolation of the relationship to η = 0 in Figure 13 appeared to increase with increase in $\bar{D}$ beyond $\bar{D}$ ≃ 0.55 mm.

A plausible interpretation of the results given in Figure 13 and the above two paragraphs is that η is increased by rainfall-induced tubulence in the flow. Such turbulence would increase the Reynolds stresses between the flow and the bed, increasing rate of sediment entrainment.

Such turbulence would be expected to increase with rainfall rate for any $\bar{D}$, and for any rainfall rate to decrease with $\bar{D}$, at least over some range of $\bar{D}$. As $\bar{D}$ increased significantly beyond the very shallow depth range investigated by Kilinc and Richardson (1973), it would be expected that η would become increasingly insensitive to rainfall rate, the character of turbulence becoming increasingly dominated by overland flow characteristics themselves.

For the soil type and slope range investigated in this series of experiments, estimates indicate that it is most unlikely that the first term in equation [33] omitted in the simplified versions considered in this section would have significantly altered the results of this analysis.

## D. The Effect of Substantial Depth of Overland Flow on η

The data discussed in this subsection were collected by Lembaga Penelitian Tanah (Centre for Soil Research) Bogor and the Institute Pertanian Bogor, Indonesia. The data were kindly communicated by Dr. Muljadi, Director of the Centre, and Mr. Suwardjo, who was responsible for the experiments, collected in the field over the period 1978–80.

The experimental plots of 14% slope and 22 m length were kept bare and exposed to natural rainfall. The soil was a latosol, and rill damage was manually repaired following any erosion event. Runoff and sediment were collected in large concrete containers so that the total sediment loss from each plot could be determined, thus yielding accurate data.

Data were collected on each event, but only mean monthly data are presented here. Up to 12 runoff events per month were recorded. η was calculated using equation [74] and total monthly data.

The equivalent ponded depth of runoff water, measured for each event, was weighted by the mass of soil eroded in that event, to give an effective mean equivalent ponded depth of runoff, denoted $\bar{D}_p$. The actual mean depth of water on the plots during an event ($\bar{D}$) would be approximately proportional to $\bar{D}_p$ in these data, since hourly rainfall rates during significant erosion events were typically on the order of 20 mm hr$^{-1}$. While more detailed rainfall rate data would be needed to calculate $\bar{D}$, Figure 14 shows a relationship of calculated η to $\bar{D}_p$, and thus approximately, to $\bar{D}$.

Despite some scatter, over much of the range of $\bar{D}_p$ η declines in an

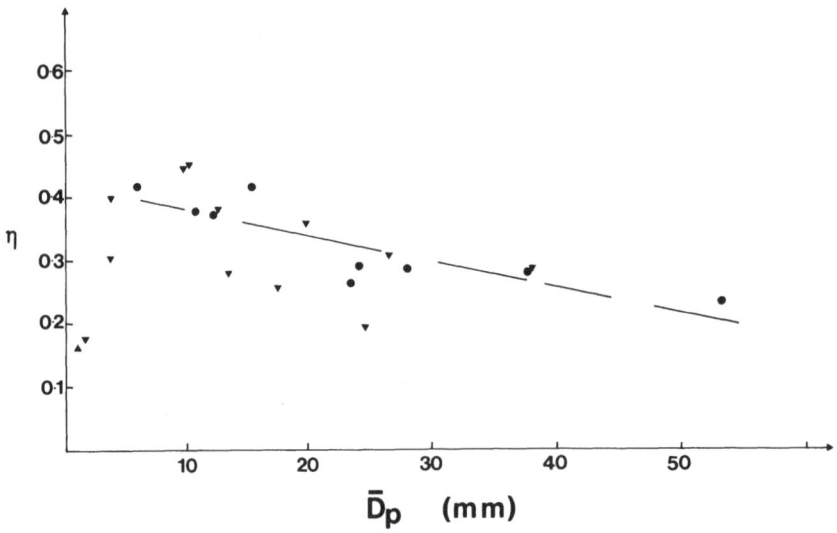

**Figure 14.** The relationship between entrainment efficiency η and equivalent ponded depth of runoff ($D_p$) for monthly data collected by the Centre for Soil Research, Bogor, Indonesia. Dashed straight line is hand-fitted to data for larger ponded depths of runoff. • represents 1978 data; ▼, 1979 data.

approximately linear fashion with increasing $\bar{D}_p$ (Figure 14). From Figure 14 η can also be seen to decline at low values of $\bar{D}_p$ ($<\sim$5 mm). The explanation for this decline follows the analysis presented in the next subsection E.

The decline in η with larger flow depths has been noted by Bagnold (1977) and others. As depth of flow increases, the mean velocity of flow becomes less influenced than it is at shallow depth by the rapidly changing velocity profile adjacent to the bed. This fact is the likely explanation for this decline in η towards values of order 0.13 typical of bed loads in streams and rivers (Allen, 1970).

A different type of explanation of the decrease in η with $D_p$ (Figure 14) is possible. As erosion continues through time with major events, it is likely that rills are increasingly cutting into more compact, less erodible material of higher shear strength. This would be expected to lead to a decrease in η, just as η was found to decrease with the firmer non-tilled soil (subsection B). However, this explanation is less likely in these experiments, since erosion damage was manually repaired following each event, thus presenting a somewhat uniform surface for erosion. At the end of the experiment, where this surface repair practice ceased but other measurements continued, $\eta$ was found to decrease to values even less than 0.1.

Over the approximately linear range of data in Figure 14, $\partial \eta / \partial \bar{D}_p = 4 \times 10^{-3} \text{mm}^{-1}$. Though the relation between equivalent ponded depth of runoff ($\bar{D}_p$) and actual mean depth ($\bar{D}$) is uncertain in these experiments, it would seem that $\partial \eta / \partial \bar{D} \approx -10 \times 10^{-3} \text{mm}^{-1}$ or $-10 \text{ m}^{-1}$. This is similar to the figure of 5 m$^{-1}$ that may be derived from the flume studies of Williams (quoted in Meynink, 1983) with a coarse sand.

## E. The Effect of Stream Power and Threshold Stream Power

Dangler *et al.* (1976) determined runoff and sediment losses under simulated rainstorms for a variety of soil types in the field of the islands of Hawaii and Oahu. The rainfall rate was 2.5 in. hr$^{-1}$ (63.5 mm hr$^{-1}$) and duration was $\sim$ 120 minutes. Two plot lengths were used, 35 ft (10.7 m), and 75 ft (22.9 m). Two successive rainstorms were applied on occasion, the first (denoted dry run) at prevailing field water content and the second (wet run) nearly 18 hrs after the first storm.

The results were analyzed in terms of soil erodibility in the USLE equation ($K$ values). The values obtained for the volcanic ash soils on the island of Hawaii for wet runs ranged from 0.07 to 0.51, and corresponding dry run values ranged from 0.08 to 0.60.

One soil type investigated was the Molokai series, a silty clay loam classified as a Typic Torrox (Oxisol). All sites were in continuous sugarcane cultivation prior to the tests. Final data from Dangler *et al.* (1976) after cessation of the storm will be used in an alternative analysis of the data for this soil in terms of the simplified theory of this section.

Because plot lengths were modest in these experiments, the form of simplified theory used is equation [71] rather than [72]. Recalling that

$$x_* = \Omega_0 / \rho g S Q, \qquad [27]$$

even if $\Omega_0$ is a constant, $x_*$ will vary substantially with time because of its dependence on $1/Q$. A time-averaged value of $x_*$, denoted $\bar{x}_*$, will be defined by

$$\bar{x}_* = \Omega_0 / \rho g S \bar{Q}, \qquad [76]$$

with

$$\bar{Q} = \int_0^{t_R} Q \, dt / t_R. \qquad [77]$$

Substituting for $\bar{x}_*$, from equation [76] into equation [71], and using equation [24], yields

$$c(L) = 2700 \, SC_r \eta \, (1 - \Omega_0/\Omega) \qquad [78]$$

$$= 2700 \, SC_r \lambda, \qquad (\text{kg m}^{-3}) \qquad [79]$$

**Table 1.** Results of Analysis of Data on Molokai-A Soil from Dangler *et al.* (1976) to Yield Values of $\lambda_1$ (for Shorter) and $\lambda_2$ (for Longer) Slope Lengths (See Text)

| Site | $L_1$ or $L_2$ | Dry run/ wet run | $\int_0^{t_R} Q\,dt$ (in.) | $t_R$ (min) | $S$ | $\lambda_1$ or $\lambda_2$ |
|---|---|---|---|---|---|---|
| 1 | $L_1$ | D | 1.59 | 120 | 0.184 | 0.094 |
|   | $L_2$ |   | 2.05 |     | 0.158 | 0.215 |
| 2 | $L_1$ | D | 2.07 | 120 | 0.055 | 0.042 |
|   | $L_2$ |   | 2.32 |     | 0.068 | 0.120 |
| 3 | $L_1$ | D | 1.73 | 120 | 0.149 | 0.136 |
|   | $L_2$ |   | 1.07 |     | 0.154 | 0.252 |
| 3 | $L_1$ | W | 4.09 | 119.5 | 0.149 | 0.123 |
|   | $L_2$ |   | 3.89 |       | 0.154 | 0.214 |
| 4 | $L_1$ | D | 0.73 | 131 | 0.145 | 0.021 |
|   | $L_2$ |   | 2.09 |     | 0.152 | 0.338 |
| 5 | $L_1$ | D | 1.79 | 125 | 0.039 | 0.104 |
|   | $L_2$ |   | 1.99 |     | 0.042 | 0.123 |
| 6 | $L_1$ | D | 1.39 | 120 | 0.051 | 0.058 |
|   | $L_2$ |   | 1.26 |     | 0.042 | 0.193 |
| 6 | $L_1$ | W | 3.25 | 120 | 0.051 | 0.079 |
|   | $L_2$ |   | 3.58 |     | 0.042 | 0.198 |
| 7 | $L_1$ | D | 2.15 | 120 | 0.081 | 0.086 |
|   | $L_2$ |   | 1.81 |     | 0.089 | 0.346 |
| 7 | $L_1$ | W | 3.85 | 120 | 0.081 | 0.126 |
|   | $L_2$ |   | 3.64 |     | 0.089 | 0.222 |

where

$$\lambda = \eta(1 - \Omega_0/\Omega). \qquad [80]$$

Since for any experiment in general neither $\eta$ nor $\Omega_0$ is known, only $\lambda$ can be calculated from data on runoff and sediment loss. Interpreting the data of Dangler *et al.* (1976) for bare Molokai-A soil using equation [79] results in Table 1, with $\lambda$ shown plotted against $\Omega$ at exit from the plane [i.e., $\Omega(L)$] in Figure 15.

There is substantial variation between plots in the value of $\lambda$, but, despite scatter presumably resulting from site-to-site variation, there is a distinct tendency for $\lambda$ to increase with stream power. This dependence of $\lambda$ on

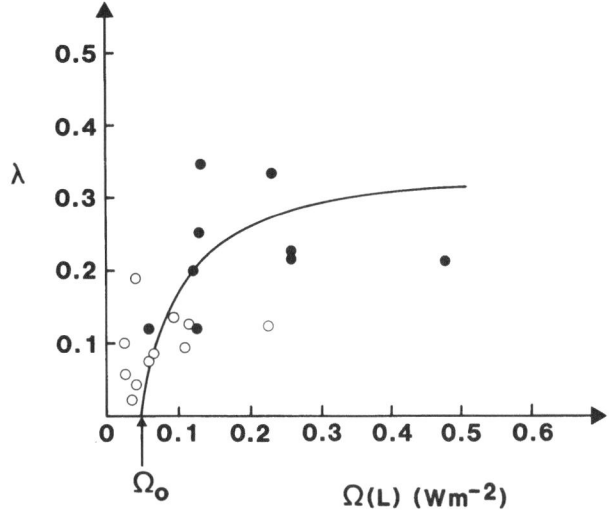

**Figure 15.** The relationship between $\lambda$ (equation [80]) and stream power $\Omega$ $(L)$ at exit from the field experimental plots of Dangler *et al.* (1976). Plots of varied slope but same soil type, exposed to simulated rainfall. Plot lengths either 10.7 m (35 ft, $\bigcirc$) or 22.9 m (75 ft, $\bullet$).

stream power, for lower values of $\Omega$, is likely to be a general characteristic. For example, it was noted in discussing Figure 14 (assuming $\eta \doteq \lambda$ in these experiments).

Though cultivation and prior crop pretreatment was the same for all plots in these experiments, there was some variation between plots in prior erosion history (Dangler *et al.*, 1976). One reason for scatter in the data (Figure 15) is likely to be variation between plots in $\Omega_0$. However, the curve shown fitted to the data in Figure 15 assumes an average value $\bar{\Omega}_0$ of $\Omega_0$, equal to 0.05 W m$^{-2}$, the value of $\Omega$ found by Loch and Donnollan (1983a) to correspond to the onset of rilling and a rapid rise in $c$ (Figure 6), though for quite different soil types.

The curve fitted to the data in Figure 15 is equation [80], assuming $\bar{\Omega}_0 = 0.05$ W m$^{-2}$ and $\eta = 0.35$. Assuming $\eta$ to be a constant independent of $\Omega$ ascribes all the dependence of $\lambda$ on $\Omega$ to the component factor $(1 - \Omega_0/\Omega)$ or $(1 - \bar{x}_*/L)$ that arises from the theory. Further detailed experimentation is required to justify the assumed constancy for a particular bare soil of the net entrainment efficiency, $\eta$. However, there is little doubt about the great importance of the factor $(1 - \Omega_0/\Omega)$ in interpreting sediment loss from small-scale experiments or from the short slope lengths typical of steep cultivated land in some third-world countries.

For bare soil, the threshold stream power $\Omega_0$ is expected to depend on soil type and its physical condition, as influenced, for example, by timing and type of cultivation. However, $\Omega_0$ may also depend substantially on the fraction $(1 - C_r)$ of the land surface with effective cover in contact with it. If so, an alternative representation of the data behind Figure 12 could be that $\eta$ is not the major variable, as is implied by the use of equation [72], but rather it is $\Omega_0$ in equation [78], which is strongly dependent on $C_r$. While further research is required to distinguish between these two alternatives, practical progress in soil conservation design can be made using the implied dependece of $\eta$ on fractional exposure $(C_r)$ shown in Figure 12.

It follows from equation [80] that $\lambda$ tends to the (assumed) constant value $\eta$ as $\Omega$ increases. An implication that follows from the form of equation [80] or Figure 15 refers to the region of significant dependence of $\lambda$ on $\Omega$ (i.e., to lower values of $\Omega$). In this region it may be shown that for the same runoff, rate of soil erosion would be expected to be higher in the presence of rills than without. In contrast, if $\lambda$ is independent of $\Omega$ for higher values of $\Omega$, then in this region the rate of soil erosion also would be independent of whatever degree of rilling may or may not take place. Thus, the importance of rilling with respect to soil loss lies particularly in the lower but still important range of $\Omega$—less than $\sim 0.5$ W m$^{-2}$ (Figure 15).

### F. Towards a Model of $\lambda$

A major objective in applying either the fuller theory of equation [33] or the simplified theory described above in subsection A is to be able to predict the likely value of $\lambda$ in any circumstance. The nature and character of the soil as influenced by its management history will have some effect on $\eta$ and probably on $\Omega_0$, so that some information related to the soil will always be needed to be able to predict $\lambda$. Whether it is most cost effective to determine $\eta$ and $\Omega_0$ by measurement of runoff and sediment loss from a plot, or in a flume, or to develop correlations so determined with other soil properties that might be more easily measured is not yet clear.

As explained by Freebairn and Rose (1982), the value of $\Omega_0$ can be readily obtained using microplots of order 1 m$^2$ subject to simulated rainfall of appropriate rate such that $x_*$ is less than the dimension $(L)$ of the microplot, and yet $x_*/L$ is significant compared with unity (equation [70]).

The finding of each labeled subsection above on the influence of different factors on $\eta$ will now be summarized in analytic form.

$$\textit{Subsection B: } \eta = b_1 \exp[-b_2(1-C_r)] \qquad [75b]$$

where $b_2 \approx 15$. The value of $b_1$ may be approximately halved by no tillage. Stream power was typically high in the experiments on which equation [75b] is based, hence neglect of term $(1 - \Omega_0/\Omega)$.

*Subsection C*: $\eta = b_3 (P - b_4)$ (Figure 13), with this dependence on $P$ only applying to shallow depths $\bar{D}$ of overland flow. The effect decays to a negligible amount for $\bar{D} >$ a few millimeters probably, this depth depending to some extent on drop size. For data interpreted using equation [75b], this effect on rainfall rate will already be included.

*Subsection D*: $\eta$ is reduced by $b_5 \bar{D}_p$ for $\bar{D}_p \gtrsim 5$ mm (Figure 14), with $b_5$ $\approx 4 \times 10^{-3}$ mm$^{-1}$ or 4 m$^{-1}$. Alternatively, and more fundamentally, this reduction can be related to the mean depth of flow $(\bar{D})$ during the runoff event, as discussed in the relevent subsection.

*Subsection E*: When stream power $\Omega$ is of such a value that $\Omega_0/\Omega$ is not negligibly small compared with unity, the data for bare soil reviewed in this subsection was not inconsistent with regarding $\eta$ a constant $(b_1)$, the variation of $\lambda$ with stream power being at least largely explained by the component term $(1 - \Omega_0/\Omega)$ from the theory (equations [78], [80]).

*Model of* $\lambda$: In a model particularly aimed at predicting the results of more important erosion events, the dependence of $\eta$ on $P$ with shallow flows will not be explicitly shown for reasons given above in discussing subsection C. With this limitation to complete generality, the above findings can be combined into a model for $\lambda$ of the following form:

$$\lambda = b_1 \exp[- b_2(1 - C_r)][1 - \Omega_0/\Omega)] - b_5 \bar{D}_p, \qquad [81]$$

where approximations have been quoted for all the constants except $b_1$. This constant represents the value of $\eta$ for bare soil $(C_r = 1)$, at substantial $\Omega$ but without $\bar{D}_p$ being too large (so $b_5 \bar{D}_p$ is small). Reviewing approximate value of $b_1$ from this section:

1. $b_1 \simeq 0.7$ for a cultivated Pellustert and Chromustert (vertisols) (Figure 12).
2. $b_1 \simeq 0.5$ for a cultivated Oxic Haplustalf (clay loam) (Loch and Donnollan, 1983a).
3. $b_1 \simeq 0.42$ for a cultivated latosol (Figure 14).
4. $b_1 \simeq 0.35$ for a cultivated Typic Torrox (Oxisol) (Figure 15).

Given adequate expansion of such experience, there seems hope that descriptive information on soil type, soil condition, and recent management history might lead to an approximate prediction of $\lambda$ of useful accuracy in any given hydrologic and management environment. That objective is certainly an aim of ongoing research.

# X. Some Brief Implications for Field Conservation Research and Conservation Planning

The aim of traditional field erosion research is to determine the effect of various management options on runoff and soil loss from plots of defined dimensions, and then to extrapolate these findings by analogy to agricultural practice. When the environment is such that significant erosion events are sporadic, such plot studies have to be maintained for long time periods, perhaps several decades, if direct use of such data is to be meaningful in the long term. The collection of such long-term data is costly, difficult to maintain, and may end up providing good data on a management option no longer relevant.

Not only is such a research methodology likely to become outdated by changing agricultural practice, but the testing of potentially useful new practices can only be achieved in the long term. When this problem of variety of conservation practices is compounded by variety in soil type, land characteristics, rainfall, and hydrologic environment, the problem can absorb a great deal of research resources, if they can be made available. Even if they are, extrapolation of research results over space and time is necessary, and traditional research methodology provides no aid to extension personnel on how this can be done.

It is in this context that a physically based model of soil erosion and deposition has been sought, and this has been described in section V, with hydrologic aspects in section II. Simplicity is a great aid to implementation, and fortunately the full theory can be simplified while retaining adequate accuracy for most purposes, as discussed in section IX. From this simplified model, the efficiency of sediment entrainment, $\eta$, emerges as the single most important parameter, apart from surface cover, $C_r$, which can be estimated directly. In section IXF current progress towards a predictive model of $\eta$ was reviewed.

Application of this simplified model to land planning at the farm scale will now be illustrated. This illustration, based on upland Australian experience (Freebairn and Rose, 1982), assumes a farm property with significant variation in slope within it, calling for conservation planning decisions on layout, land use, and conservation management options. The illustration assumes that the relationship between $\eta$ and $C_r$ has been experimentally determined and the relationship of Figure 12 will be used. It follows from this figure and equation [72] that there is a relationship between $c$, land slope $S$, and the fraction $(1-C_r)$ of soil covered; this relationship is shown in Figure 16.

To illustrate use of Figure 16 in farm planning, suppose annual runoff with a bare fallow treatment in a particular environment is 60 mm yr$^{-1}$. It is desired to find the maximum slope limitation for this management practice if a certain "tolerable" rate of soil loss (or "$T$-value") is adopted for these particular soil and climate conditions.

There is need to better establish what such $T$-values may be, but for illustration a figure of 10 Mg (or tonne) ha$^{-1}$yr$^{-1}$ (or 1 kg m$^{-2}$ yr$^{-1}$) will be adopted. From equation [1], with $c$ constant (equation [72]),

$$\sum q_s = c \sum q, \qquad [82]$$

where the summation is over time. With both $\sum q_s$ and $\sum q$ specified as above, then from equation [82] a particular value of $c$ is implied. With $\sum q = 0.06$ m yr$^{-1}$, and $\sum q_s = 1$ kg m$^{-2}$ yr$^{-1}$, $c = 16.7$ kg m$^{-3}$ is obtained.

**Figure 16.** Graph allowing sediment concentration $c$ (kg m$^{-3}$) to be read off as a function of land slope ($S$ expressed in percent) and fraction of soil cover $(1-C_r)$. Based on the relation shown in Figure 12 and approximate equation [72] in the text. Percentages for sediment concentration are approximate only (Rose *et al.*, 1983a).

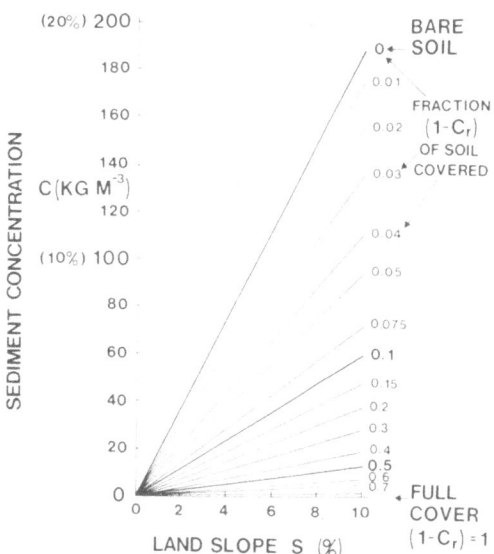

Entering the ordinate of Figure 16 with $c = 16.7$ kg m$^{-3}$, we see that if soil is bare the limiting land slope to meet the stated $T$-value is approximately 1%. Continuing with the same value of $c$ in Figure 16, we see that if a 10% cover [i.e., $(1-C_r = 0.1)$] could be maintained then the limiting slope rises to about 3%, and to 6% for a 30% soil cover.

However, this simple extrapolation to higher covers leads to conservative practice because in general a higher surface cover reduces runoff. In the example under consideration (Freebairn and Rose, 1982), incorporation of stubble would reduce $\sum q$ to some 39 mm yr$^{-1}$, so that now $c = 25.6$ kg m$^{-3}$. Incorporated stubble provides cover $(1-C_r) = \sim 0.2$, so that a slope of some 6.5% would still allow the $T$-value to be met. This may be compared with an inferred land slope limitation of some 4.5% if the effect of cover on hydrology were ignored (Figure 16).

This illustrates the intimate links between surface management, surface hydrology, and sediment transport. However, it also illustrates a methodology whereby these links can be rather simply dealt with, provided the adequate basic data is determined.

## Acknowledgments

Some of this work was supported by grants from the Rural Credits Development Fund of the Reserve Bank of Australia. Valuable discussion with the Queensland Department of Primary Industries soil conservation staff, and with D.M. Freebairn in particular, is gratefully acknowledged. Thanks are due to Mr T. Proffitt and Mr J. van der Molen for checking calculations.

**Appendix: List of Major Symbols**

| Greek/ Roman symbol | Description | Defining equation number (if applicable) |
|---|---|---|
| $a$ | Detachability of the soil by rainfall | [20] |
| $c$ | Sediment concentration | [1] |
| $c(L)$ | Sediment concentration at $x = L$ | |
| $C_e$ | Fraction of soil surface unprotected from raindrop detachment | [20] |
| $C_r$ | Fraction of soil surface unprotected from entrainment by overland flow | |
| $d$ | Sediment deposition rate | [22] |
| $D$ | Analytic approximation to depth of overland flow | |
| $\bar{D}$ | Average value of depth $D$ on plane | |
| $D_c$ | Erosion capacity of rill flow | |
| $D_r$ | Rill erosion rate | |
| $e$ | Rainfall detachment rate | [20] |
| $g$ | Acceleration caused by gravity | |
| $i$ as a subscript | Refers to a particular sediment size range | |
| $I$ | Number of sediment size ranges, infiltration rate | |
| $K$ | $0.276\eta$ | [27] |
| $K_1$ | Hydraulic coefficient | [5] |
| $L$ | Length of plane | |
| $m$ | Net flux of entrained sediment, expressed as mass per unit width of stream bed; hydraulic coefficient | [25] [4] |
| $M_s$ | Accumulated mass of sediment leaving the plane of width $W$ at $x = L$ | [35] |
| $n$ | Manning's roughness coefficient | |
| $p$ | Power of $P$ in equation for rainfall detachment rate; rate of sediment transport from plane to channel | [20] [47] |
| $P$ | Rainfall rate | |
| $q$ | Volumetric water flux per unit width of plane | [1] |

**Appendix: List of Major Symbols** *(continued)*

| Greek/ Roman symbol | Description | Defining equation number (if applicable) |
|---|---|---|
| $q(L)$ | Value of $q$ at $x = L$, the bottom of the plane | |
| $q_s$ | Sediment flux per unit width of plane | [1] |
| $q_s(L)$ | Value of $q_s$ at $x = L$ | [34] |
| $Q$ | Runoff rate per unit plane area | |
| $r$ | Sediment entrainment rate | [29] |
| $R$ | Excess rainfall rate | [2] |
| $S$ | Slope of the plane (the sine of the angle of land surface inclination) | |
| $t$ | Time | |
| $t_R$ | Duration of runoff event | |
| $T_c$ | Flow transport capacity | [19] |
| $v_i$ | Settling velocity of sedimentary particles of size range $i$ | [22] |
| $w$ | Immersed weight of sediment passing unit width of stream bed per unit time | [23] |
| $W$ | Width of plane | |
| $x$ | Distance down-slope from the top of the plane | |
| $x_*$ | Value of $x$ beyond which $r > 0$ | [27] |
| $\bar{x}_*$ | Time average value of $x_*$ | [76] |
| $x_p$ | Distance downplane to junction of two branches of exact hydrologic equation solution | |
| $\gamma_i$ | $(1 + v_i/Q)$ | [31] |
| $\eta$ | Efficiency of entrainment by overland flow ($0 \leq \eta \leq 1$) | [23] |
| $\rho$ | Density of water | |
| $\sigma$ | Density of sedimentary material | |
| $\Omega$ | Stream power | [23], [24] |
| $\Omega_0$ | Threshold value of $\Omega$ | [23] |

# References

Allen, J.R. 1970. *Physical processes of sedimentation.* George Allen and Unwin, London. 248 pp.

Alonso, C.V., W.H. Neibling, and G.R. Foster. 1981. Estimating sediment transport capacity in watershed modelling. *Trans. Am. Soc. Agr. Eng.* 24(5):1211–1220, 1226.

Anonymous. 1943. *A study of new methods for size analysis of suspended sediment samples.* Report No. 7: Office of Indian Affairs, Bureau of Reclamation T.V.A., Corps of Eng. Geol. Surv., Dept. of Agr. and Iowa Institute of Hydraulic Res., University of Iowa.

Bagnold, R.A. 1956. The flow of cohesionless grains in fluids. *Phil. Trans. Roy. Soc. London A* 249:235–297.

Bagnold, R.A. 1966. An approach to the sediment transport problem from general physics. *U.S. Geol. Surv. Prof. Paper* No. 442-I.

Bagnold, R.A. 1977. Bed load transport by natural rivers. *Water Resour. Res.* 13:303–311.

Campbell, S.Y., and J.Y. Parlange. 1984. Overland flow on converging and diverging surfaces—assessment of numerical schemes. *J. Hydrol.* 70:265–275.

Campbell, S.Y., J.Y. Parlange, and C.W. Rose. 1984. Overland flow on converging and diverging surfaces—kinematic model and similarity solutions. *J. Hydrol.* 67:367–374.

Dangler, E.W., S.A. El-Swaify, L.R. Ahuja, and A.P. Barnett. 1976. Erodibility of selected Hawaii soils by rainfall simulation. *USDA ARS and Univ. Hawaii Agric. Expt. Sta. Publ.* ARS-35. 113 pp.

El-Swaify, S.A., E.W. Dangler, and C.L. Armstrong. 1982. *Soil erosion by water in the tropics.* College of Tropical Agriculture and Human Resources. Univ. of Hawaii. Res. Extension Series 024. 173 pp.

Forrester, J.E. 1970. *Industrial dynamics.* MIT Press, Cambridge, MA.

Foster, G.R. 1982. Modelling the erosion process. In: C.T. Haan (ed.), *Hydrologic modelling of small watersheds. Am. Soc. Agr. Eng. Monogr.* No. 5. pp. 297–379. St. Joseph, MI.

Foster, G.R., and L.J. Lane. 1981. Simulation of erosion and sediment field from field-sized areas. In: R. Lal and E.W. Russell (eds.), *Tropical agricultural hydrology.* pp. 375–394. John Wiley and Sons, New York.

Foster, G.R., and L.D. Meyer. 1972. A closed form soil erosion equation of upland areas. In: H.W. Shen (ed.), *Sedimentation symposium to honor Professor H.A. Einstein.* pp. 12.1–12.19. Fort Collins, CO.

Foster, G.R., and L.D. Meyer. 1975. Mathematical simulation of upland erosion by fundamental erosion mechanics. In: *Present and prospective technology for predicting sediment yields and sources.* pp. 190–207. USDA Res. Service, Southern Region, ARS-S-40.

Foster, G.R., L.D. Meyer, and C.A. Onstad. 1977. An equation derived from basic erosion principles. *Trans. Am. Soc. Agr. Eng.* 20(4):678–682.

Freebairn, D.M., and W.C. Boughton. 1981. Surface runoff experiments on the Eastern Darling Downs. *Austr. J. Soil Res.* 19:133–146.

Freebairn, D.M., and C.W. Rose. 1982. *Integration and extrapolation of field erosion research using a physically based erosion/deposition model.* Department of Primary Industries, Queensland. Report Q082014.

Gumbs, F.A., and J.I. Lindsay. 1982. Runoff and soil loss in Trinidad under different crops and soil management. *Soil. Sci. Soc. Am. J.* 4:1264–1266.

Hillel, D. 1980. *Applications of soil physics.* Academic Press, New York. 385 pp.

Kilinc, M., and E.V. Richardson. 1973. Mechanics of soil erosion from overland flow generated by simulated rainfall. Colorado State University, Ft. Collins, CO. *Hydrology Papers* No. 63.

Kinnell, P.I.A. 1982. Initial results from runoff and soil loss plots at Ginninderra, A.C.T. *CSIRO Division of Soils, Canberra, Div. Report* No. 61.

Knisel, W.G. (ed.) 1980. CREAMS: A field-scale model for chemicals, runoff, and erosion from Agricultural Management Systems. *USDA Conservation Research Report* No. 26. 643 pp.

Kynch, G.J. 1952. A theory of sedimentation. *Trans. Faraday Soc.* 48:166–176.

Lal, R. 1982. Effects of slope length and terracing on runoff and erosion on a tropical soil. In: *Proc. of the Exeter symposium on recent developments in the explanation and prediction of erosion and sediment yield.* pp. 23–31 IAHS Publ. No. 137.

Lane, L.J., and E.D. Shirley. 1982. Modelling erosion in overland flow. In estimating erosion and sediment yield on rangelands. *USDA, ARS. Agr. Reviews and Manuals.* ARM-W-26/June 1982.

Lane, L.J., and D.A. Woolhiser. 1977. Simplifications of watershed geometry affecting simulation of surface runoff. *J. Hydrol.* 35:173–190.

Loch, R.J., and T.E. Donnollan. 1983a. Field rainfall simulator studies on two clay soils of the Darling Downs, Queensland. I. The effect of plot length and tillage orientation on erosion processes and runoff and erosion rates. *Austr. J. Soil Res.* 21:33–46.

Loch, R.J., and T.E. Donnollan. 1983b. Field rainfall simulator studies on two clay soils of the Darling Downs, Queensland. II. Aggregate breakdown, sediment properties and soil erodibility. *Austr. J. Soil Res.* 21:47–58.

Meynink, W.J.C. 1983. C.R.C.E. Sediment transport model—development, calibration, verification. *Catchment River and Coastal Eng. Pty. Ltd. Tech. Report* No. 83/1. 59 Cairns Tce, Red Hill, Qld.

Moss, A.J. 1972. Bed-load sediments. *Sedimentology* 18:159–219.

Moss, A.J. 1979. Thin-flow transportation of solids in arid and non-arid areas: a comparison. In: *Symposium on the hydrology of areas of low precipitation.* pp. 435–445. IAHS Publ. No. 128.

Moss, A.J., and P. Green. 1983. Movement of solids in air and water by raindrop impact. Effects of drop-size and water-depth variations. *Austr. J. Soil Res.* 21:257–269.

Moss, A.J., and P.H. Walker. 1978. Particle transport by continental water flows in relation to erosion, deposition, soils and human activities. *Sed. Geol.* 20:81–139.

Moss, A.J., P. Green, and J. Hutka. 1982. Small channels: their experimental formation, nature and significance. *Earth Surf. Proc. Land.* 1:401–415.

Moss, A.J., P.H. Walker, and J. Hutka. 1979. Raindrop-stimulated transportation in shallow water flows: an experimental study. *Sed. Geol.* 22:165–184.

Moss, A.J., P.H. Walker, and J. Hutka. 1980. Movement of loose, sandy detritus by shallow water flows: an experimental study. *Sed. Geol.* 25:43–66.

Palmer, R.S. 1964. The influence of a thin water layer on water drop impact forces. *Int. Assoc. Sci. Hydrol. Publ.* 65:141–148.

Parlange, J.-Y., C.W. Rose, and G. Sander. 1981. Kinematic flow approximation of runoff on a plane: an exact analytical solution. *J. Hydrol.* 52:171–176.

Rose, C.W. 1960. Soil detachment caused by rainfall. *Soil Sci.* 89:28–35.

Rose, C.W. 1961. Rainfall and soil structure. *Soil Sci.* 91:49–54.

Rose, C.W. 1984. Recent advances in research soil erosion processes. In: *Proc. National Soils Conference*, Brisbane, Australia. pp. 2–26. Austr. Soc. Soil. Sci. Inc.

Rose, C.W., D.M. Freebairn, and G.C. Sander. In press. *GNFIL: A Griffith University program for computing infiltration from field hydrologic data*. School of Australian Environmental Studies Monograph, Griffith University, Brisbane.

Rose, C.W., B.R. Roberts, and D.M. Freebairn. 1983a. Soil conservation policy and a model of soil erosion. In: D.E. Byth, M.A. Foale, V.E. Mungomery, and E.S. Wallis (eds.), *New technology in field crop protection*. pp. 212–226. Austr. Inst. Agr. Sci., Melbourne.

Rose, C.W., J.-Y. Parlange, G.C. Sander, S.Y. Campbell, and D.A. Barry. 1983b. A kinematic flow approximation to runoff on a plane; an approximate analytic solution. *J. Hydrol.* 62:363–369.

Rose, C.W., J.R. Williams, G.C. Sander, and D.A. Barry. 1983c. A mathematical model of soil erosion and deposition processes. I. Theory for a plane land element. *Soil Sci. Soc. Am. J.* 47:991–995.

Rose, C.W., J.R. Williams, G.C. Sander, and D.A. Barry. 1983d. A mathematical model of soil erosion and deposition processes. II. Application to data from an arid-zone catchment. *Soil Sci. Soc. Am. J.* 47:996–1000.

Schroeder, S.A., G.R. Foster, W.C. Moldenhauer, and J.V. Mannering. 1982. Hydraulic conductivity of soil as determined from cumulative runoff. *Soil Sci. Soc. Am. J.* 46:1267–1270.

Singer, M.J., and P.H. Walker. 1983. Rainfall-runoff in soil erosion with simulated rainfall, overland flow and cover. *Austr. J. Soil Res.* 21:109–122.

Walker, P.H., P.I.A. Kinnell, and P. Green. 1978. Transport of a noncohesive sandy mixture in rainfall and runoff experiments. *Soil Sci. Soc. Am. J.* 42:793–801.

Walker, P.H., J. Hutka, A.J. Moss, and P.I.A. Kinnell. 1977. Use of a versatile experimental system for soil erosion studies. *Soil Sci. Soc. Am. J.* 41:610–612.

Wischmeier, W.H., and D.D. Smith. 1978. *Predicting rainfall erosion losses—a guide to conservation planning.* Agr. Handb. No. 537, USDA. Govt. Printing Office, Washington, DC.

Woolhiser, D.A., and J.A. Liggett. 1967. Unsteady one-dimensional flow over a plane—the rising hydrograph. *Water Resour. Res.* 3:753–771.

# Dynamics and Availability of Major Nutrients in Soils

## Konrad Mengel[*]

## I. Introduction

In his book *The Organic Chemistry and Its Application on Agriculture and Physiology* Liebig (1841) stated: "Als Prinzip des Ackerbaues muß angesehen werden, daß der Boden in vollem Maße wieder erhalten muß,

---

[*]Institut für Pflazenernöhrung, D-6300, Südanlage 6, Federal Republic of Germany.

was ihm genommen wird" (it must be borne in mind that as a principle of arable farming, what is taken from the soil must be returned to it in full measure). This was the theoretical basis of a development with far-reaching consequences. According to this principle, fertility of many soils could be maintained or even restored. The efficiency of crop production was tremendously improved, as evidenced by the yield attained per unit of land. At the time Liebig began his studies in Germany 1 ha of cultivated land produced food for about one person. Today's figure is 4.5 people. During this development, fertilizer use also increased considerably. Future crop production, whether in countries with a highly developed or less developed agriculture, will be based on the same principle as that stated by Liebig: maintainance and restoration of soil fertility by substituting the plant nutrients exported with crops from the field or lost by other processes by means of fertilizer application. Besides this, as already has been done in the past, new cultivars should be bred with a higher nutrient efficiency (Mengel, 1983). This aspect, however should not be over-estimated, since a heavy crop, e.g., wheat (*Triticum aestivum*), stores 200 kg N and 40 kg P ha$^{-1}$ in its grain. These nutrients are exported from the field and must be replenished. It thus appears that as the intensity of crop production increases, the importance of fertilizer also increases (Arnon, 1969).

Commercial fertilizers draw on limited resources such as energy and raw materials, and it is for this reason that the efficient use of fertilizers is a challenge for those who are involved in fertilization. Soil scientists in particular are confronted with the need to avoid fertilizer losses and to adjust fertilizer rates to soil and crop requirements. To meet this challenge, more knowledge is necessary about plant nutrient behavior in the soil and about a phenomenon that generally is called nutrient availability, a term that at present is poorly defined.

Nutrient availability can be defined in both physico-chemical and biological terms. In a physico-chemical sense, a nutrient is available if it is dissolved in the soil solution or if it can be easily dissolved or desorbed and thus replenish the soil solution. Such a nutrient, however, will only be taken up by the crop if it is reached by a plant root. A soluble phosphate, available from the physico-chemical status but located in a hard clod, that cannot be penetrated by plant roots is unavailable in a biological sense. Both the physico-chemical and biological availability aspects will be considered in the following sections.

The most common plant nutrients included in fertilizers are nitrogen, phosphorus, and potassium, and these are the only plant nutrients treated in this contribution. The topic already represents a vast field, and space does not permit all aspects to be considered nor all publications worthy of quotation to be cited. The reviewer's intention in this paper is to contribute to improved fertilizer efficiency.

## II. Factors and Processes of Plant Nutrient Availability

### A. Soil Factors and Processes

Numerous solution culture experiments have shown the close relationship between the nutrient concentration in the soil solution and the rate of nutrient uptake by the plant. In a low concentration range $< 0.5$ mmol $L^{-1}$, this relationship is almost linear for the uptake of $NO_3^-$, $NH_4^+$, $K^+$, $H_2PO_4^-$, and $HPO_4^{2-}$ and can be described by the equation

$$U = 2\pi r\, \alpha \cdot C_r, \qquad [1]$$

where $U$ is the uptake of a root segment of 1 m, $r$ is the root radius, $C_r$ is the concentration of the ion at the root surface, and $\alpha$ is the root absorbing power.

The root absorbing power depends on the energetic status of roots. A high absorbing power means that a high proportion of ions impinging on the root surface is absorbed and vice versa. The term $\alpha$ is therefore not a constant but changes with metabolic conditions. Ion concentration at the root surface $(C_r)$ depends on $\alpha$, since a high absorbing power tends to decrease $C_r$; ion concentration at the root surface depends also on the rate of movement of the ion from the bulk soil towards the root. This movement may be brought about by diffusion and/or mass flow. If the rate of movement of a particular ion species is high in comparison with the rate of ion uptake by the root, the concentration of the particular ion species at the root surface will not decline, but even an ion accumulation may occur, as has been shown for $Ca^{2+}$ by Barber (1974). This may also be the case for other ion species with high concentrations in the soil solution, such as $Mg^{2+}$ and $Na^+$ and $Cl^-$ in saline soils. Nitrate concentration of the soil solution may also be high, and thus a nitrate accumulation at the root surface is possible.

On the other hand, ion species that are generally in low concentration in the soil solution but are taken up by the root at high rates become depleted at the root surface. This is particularly true for phosphate and also applies for $K^+$ and $NH_4^+$. The depletion of these ions has been shown by various authors: for phosphate by Lewis and Quirk (1967), Bhat and Nye (1974), and Hendriks et al. (1981); for $K^+$ or $Rb^+$ by Barber (1962) and Hendriks et al. (1981); and for $NH_4^+$ by Liu Zhi-Yu and Qin Sheng-Wu (1981). Ion depletion at the root surface means creating a sink for the ion species in question. The concentration gradient established between the root surface and the bulk soil solution provides the condition for a net thermal movement of the ion in question. It is generally believed that phosphate, $K^+$, and $NH_4^+$ are translocated mainly along this concentration gradient and hence diffusion is the principal transport mechanism for them.

Mass flow is assumed to be the major mechanism for moving nitrate towards the roots (Renger and Strebel, 1976). Liao and Bartholomew (1977) found a close correlation between amounts of water transpired and $NO_3^-$ taken up by corn (*Zea mays*). Plants, however, also take up $NO_3^-$ at high rates under low transpiration conditions, which means a low flow rate of water in the soil towards the roots. Under such conditions, $NO_3^-$ diffusion may play a major role. This has been shown by Strebel et al. (1980) in a field experiment of a Udalf soil derived from loess. During the early development of spring wheat, about 50% of the $^{15}N$-labeled fertilizer $NO_3^-$ was translocated by mass flow, but with an advance in crop development this percentage dropped, and in the later stage more nitrate was translocated by diffusion than by mass flow. A similar result was obtained by Strebel et al. (1983) when studying nitrate and water flow under sugar beets in a field experiment. The most important results of this investigation are shown in Figure 1, which depicts N uptake of the crop from four different soil layers and the percent proportion of the absorbed N translocated to the roots by diffusion. Nitrogen uptake from the upper soil layer (0 to 0.3 m) showed a peak in June. The N uptake peaks for the deeper soil layers shifted to later dates as the depth of layers increased. It is therefore obvious that as root length increased, the soil layers were mined

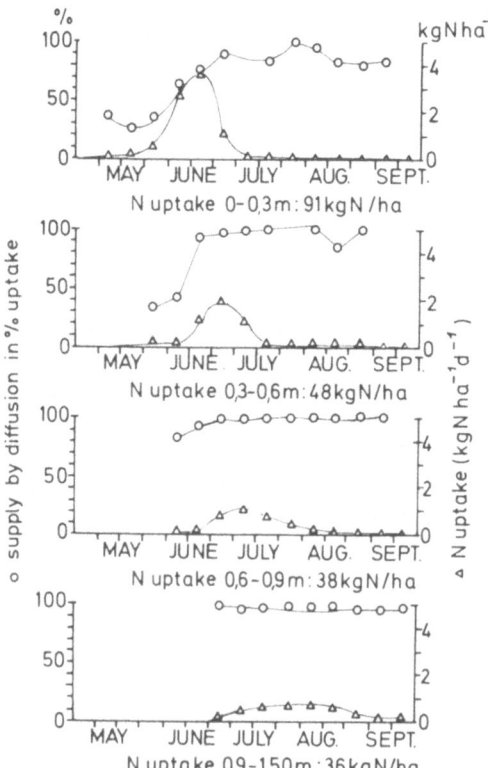

**Figure 1.** Uptake rates of nitrate (kg N $ha^{-1}d^{-1}$) of sugar beets and percentage of nitrate supplied by diffusion. After Strebel et al. (1983), *Mitt. Dtsch. Bodenk. Ges.* 38:153–158. Copyright © 1983 by Deutsche Bodenkundliche Gesellschaft. Used with permission.

for nitrate. In the upper soil layer a high $NO_3^-$ concentration prevailed at the beginning of the growing period, while the nitrate uptake rate of the crop was low. Such conditions favor mass flow transport. With an increase of N uptake in June, the proportion of mass flow diminished, and in the later stage almost 100% of the N absorbed by the crop was translocated by diffusion. In the deeper soil layers, diffusion was by far the most important process because of the low nitrate concentrations. These examples may show that diffusion in soil is also of relevance for $NO_3^-$.

Diffusion is described by Fick's first law:

$$F = -D \, (dC/dx) \qquad [2]$$

where $F$ is the flux and $dC/dx$ the concentration gradient. $D$ is the diffusion coefficient that generally describes the diffusivity of a homogeneous medium. Since the soil is not homogeneous, the concept of a diffusive flux in the soil medium presents difficulties. Following Nye's (1979) proposal, the diffusion coefficient for a soil medium is represented by the formula

$$D = D_e \, \theta \, f_e \, (dC_e/dC) + D_E, \qquad [3]$$

where D is the diffusion coefficient in the whole soil medium, $D_e$ is the diffusion coefficient in the free water of the soil, $\theta$ is the fraction of the soil volume filled with solution, $f_e$ is the impedance factor, $C_e$ is the ion concentration in the soil solution, $C$ equals ion concentration in the whole soil medium, and $D_E$ is the term for surface diffusion.

$D_e$ of the soil solution is in the range of the diffusion coefficients for ions dissolved in water, which is about 0.5 to 2.0 · $10^{-9}$ m$^{-2}$ s$^{-1}$. The term $\theta$ represents the water cross section for diffusion; it is high in water-saturated soils and low in dry soils and is thus of crucial importance for ion diffusion in soil. The impedance factor $(f_e)$ is related to the tortuosity, which means the passes and bypasses an ion has to follow on its way towards the roots. The term $f_e$ also includes the reduced mobility of ions in adsorbed water layers and the exclusion of anions from narrow pores. Both terms $\theta$ and $f_e$ decrease considerably with a decrease in soil moisture. According to Nye (1979), the product $\theta \times f_e$ may decrease by a factor of 100 when soil moisture decreases from $-10$ to $-10,000$ kPa. This shows that in very dry soils, ion diffusion is almost zero (Rowell et al., 1967). At such low soil water potential plants may suffer from water stress, so nutrient supply will play only a secondary role. In cases where plants may utilize water from deeper soil layers, however, impeded nutrient diffusion in the upper soil layer may result in an insufficient nutrient supply. Mengel and Casper (1980) found a highly significant correlation between the N uptake of corn and soil moisture with a model experiment, in which a split root technique was used to provide the plant with roots in both dry and wet soil. A clear relationship between soil moisture and N uptake of the crop has also been found under field conditions (Smika et al., 1965; Young et al., 1967). Under dry soil conditions $NO_3^-$ may accumulate in the top layer of the soil

(Page and Talibudeen, 1977). The negative influence of low soil moisture on the uptake of $K^+$ by plants was reported by Mederski et al. (1960) and by Mengel and von Braunschweig (1972).

The term $dC_e/dC$ in equation [3] indicates that the diffusion coefficient in the soil medium depends also on the concentration of the ion in the soil solution $(C_e)$ and on the concentration of this ion species in the total soil system $(C)$. This latter term actually relates to the ion fraction of the soil that is equilibrated with the ions in solutions. Thus, $C$ stands for quantity and $C_e$ for the intensity. For phosphate, $C$ is identical to the concentration of isotopically exchangeable phosphate; for $K^+$ and $NH_4^+$ $C$ denotes the concentration of exchangeable and to some degree also the concentration of non-exchangeable $K^+$ or $NH_4^+$, respectively. A low numerical value of $dC_e/dC$ means that the diffusion of an ion species is hampered by adsorption to soil particles and vice versa. The ratio $dC_e/dC$ is the reciprocal to the buffer power:

$$b = dc/dC_e. \qquad [4]$$

Substituting $b$ for $dC_e/dC$, the diffusive flux $(J)$ can be described by the equation

$$J = - \frac{D_e \, \theta \, f_e}{b} \cdot \frac{dC_e}{dx}, \qquad [5]$$

where $b$ denotes the buffer power of the ion in question. In equation [5] the term $D_E$ (see equation [3]) has been neglected, since only in special cases $D_E$ may play some role in nutrient diffusion in the soil, as will be shown below. From equation [5] it is evident that $dC_e$ is a further decisive factor of nutrient diffusion in the soil. In the plant root/soil system $dC_e$ stands for the concentration gradient between the ion concentration in the bulk soil solution and the ion concentration at the root surface. The concentration gradient is reflected by the depletion profile being established when plant roots absorb nutrients at a higher rate than the rate of nutrient translocation towards the roots (Lewis and Quirk, 1967; Drew et al., 1969).

In Figure 2 such a depletion profile is shown for phosphate from the work of Hendriks et al. (1981). The concentration gradient of the depletion profile depends on the concentration of the ion species in the bulk soil solution and on the degree to which this is lowered at the root surface by plant uptake. For $K^+$ and phosphate very low concentrations in the range of 2.0 μmol $L^{-1}$ may be attained at the root surface (Hendriks et al., 1981; Claassen and Barber, 1976; Claassen et al., 1981). The bulk concentration of a plant nutrient can be increased by fertilizer application, and thus the diffusive flux of nutrients towards the roots can be improved.

A plant root growing into a soil generally will encounter at first a relatively high nutrient concentration (phosphate, $K^+$, $NH_4^+$), in the range of the concentration of the bulk soil solution. With the uptake of nutrients,

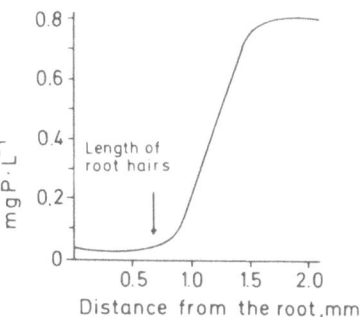

**Figure 2.** Depletion of phosphate in the soil solution around a corn root. After Hendriks *et al.* (1981), *Z. Pflanzenern. Bodenk.* 144:486–499. Copyright © 1981 by Verlag Chemie GmbH. Used with permission.

a nutrient depletion at the root surface occurs; the associated depletion profile becomes flatter with an advance in nutrient uptake (Hendriks *et al.*, 1981; Lewis and Quirk, 1967; Claassen *et al.*, 1981). Figure 3 shows this change of the depletion profile for $NH_4^+$ of rice (*Oryza sativa*) roots investigated in a typical paddy soil by Liu Zhi-yu and Qin Sheng-Wu (1981). Nutrient depletion around the root caused by the diffusive flux of nutrients towards the root surface is not only related to the nutrient intensity ($C_e$) but also to the nutrient quantity. The dissolved ions diffusing towards the roots are partially replenished by desorption. This process is of particular importance for phosphate, $K^+$, and $NH_4^+$. The replenishment of nutrients is reflected by the buffer power for a particular nutrient (see formula [4]). A high buffer power means that the unit of ions in solution (concentration) is equilibrated with a high amount of desorbable ions of the same species. Plotting the amount of desorbable (exchangeable) ions on the y axis and the corresponding ion concentration of the equilibrated solution on the x axis results in a curve whose steepness represents the buffer power. In soils well buffered for a particular ion, the concentration in the soil solution decreases only at low rates when the ion is taken up by roots, owing to the fact that ions removed from the solution are largely

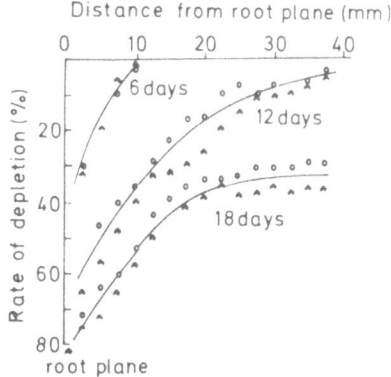

**Figure 3.** Depletion of fertilizer $^{15}NH_4^+$ at the root surface of rice measured 6, 12, and 18 days after fertilizer application. ○ and △ denote measurements on opposite soil blocks. After Liu Zhi-yu and Qin Sheng-wu (1981). Used with the permission of the publisher and the authors.

replenished by desorption (Claassen *et al.*, 1981). Hence the mean concentration at the root surface is generally higher for a well-buffered soil in comparison with a poorly buffered soil. The importance of the buffer power becomes evident if the duration of ion uptake is integrated into the uptake formula (Drew *et al.*, 1969)

$$U_t = 2\pi r \cdot a \cdot \bar{C_r} t, \qquad [6]$$

where $U_t$ is the absorbed quantity during a given time $t$, $\bar{C_r}$ is the mean ion concentration at the root surface, $t$ in time of uptake, and $\bar{C_r}$ is the mean of the initial concentration and the concentration after a given time. The latter concentration will be the lower the poorer the buffer power. Hence in poorly buffered soils the $C_r$ term will also be low. From this it follows that soils with a low buffer power for a particular nutrient require a high initial concentration of the ion in question and vice versa. This was actually found by Mengel and Busch (1982) when studying the $K^+$ uptake of ryegrass in relation to the $K^+$ concentration in the bulk soil solution and the K buffer power of nine soils widely differing in chemical and physical properties. The importance of the phosphate buffer power for the P uptake has been evidenced by Olsen and Watanabe (1970), Holford (1976), and Nair and Mengel (1984). In the model for $K^+$ uptake established by Claassen and Barber (1976), the buffer power is an essential factor. Recent results of Silberbush and Barber (1983b, 1983c) also emphasize the importance of the buffer power for $K^+$ and phosphate.

According to Barber *et al.* (1963), interception plays a negligible role in the process of nutrient exploitation of the soil. The question, however, is not yet clarified regarding the extent that plant nutrients may reach the root by surface migration (diffusion). Surface diffusion, as indicated with $D_E$ in formula [3], has been observed in clay pastes. Ions may migrate on outer and even inner surfaces. Thus, according to Nye (1979) the interlayer ions in fully expanded 2:1 clay minerals may be nearly as mobile as the ions on external surfaces. Surface migration does not occur in normal soils because of breaks in the continuous pathway consisting of adsorbed water molecules (Nye, 1979). It is, however, feasible that clay particles near the root surface may be in close contact with roots and/or root hairs. Slime exuded by the root may help to form a continuous layer of adsorbed water molecules from the interior of the clay mineral to the root surface (Matar *et al.*, 1967). Thus, Breisch *et al.* (1975) have shown by means of electron micrographs that the mucilage secreted by plant roots can adsorb clay minerals, providing a pathway that would allow a migration of cations from the clay to the root, the latter acting as a strong sink particularly for $K^+$ and $NH_4^+$. Claassen and Jungk (1982) found that roots of corn depleted soil $K^+$ in a zone of about 2 mm distance from the root surface to a very low level. From this zone not only the exchangeable $K^+$ but also considerable amounts of non-exchangeable $K^+$ had been removed by the plant. The $K^+$ concentration near the root surface was 3 µmol $L^{-1}$, and it is suggested that at this low level a net release of $K^+$ from interlayer $K^+$ occurs. This finding is in good agreement with laboratory experiments of Martin and Sparks (1983), who found a net release of non-exchangeable $K^+$ if the $K^+$ concentration of the contact solution was about 1 µmol $L^{-1}$.

Basically $NH_4^+$ release from interlayer position will follow the same mechanism: A net release will occur if the contact solution is extremely low in its $NH_4^+$ concentration. Because most plant species can take up $NH_4^+$ at high rates (Breteler and Smith, 1974; Mengel and Viro, 1978), the soil solution near the root surface can be quickly depleted of $NH_4^+$ and thus suitable conditions for a net release of $NH_4^+$ are provided. According to this concept, interlayer $NH_4^+$ and $K^+$ of vermiculite micas and illites will be released to a major extent only if they are in the depletion zone of roots. This assumption is supported by the finding that the release of non-exchangeable $NH_4^+$ follows the pattern of root growth. Thus, the curves in Figure 7 show that the release of non-exchangeable $NH_4^+$ in the deeper soil layer (0.6 to 0.9 m) occurred at a later stage when the roots had reached the deeper zone (Mengel and Scherer, 1981).

The interlayer position from which $K^+$ or $NH_4^+$ were desorbed may be occupied by other cation species, such as $Ca^{2+}$, $Mg^{2+}$, and $Na^+$ generally abundantly present near the root surface. This is especially true for $Ca^{2+}$ (Barber, 1974). The interlayer exchange of $K^+$ or $NH_4^+$ for $Ca^{2+}$ leads to an expansion of the clay mineral that in addition will contribute to the release of so-called non-exchangeable $NH_4^+$ and $K^+$. Diffusion coefficients for $K^+$ and $NH_4^+$ in micas and vermiculites are on the order of $10^{-19}$ $m^{-2}$ $s^{-1}$. Although these values are low, plants may feed sufficiently from interlayer $K^+$ for a longer time under favorable release conditions (Steffens and Mengel, 1979).

Diffusion in clay interlayers and on clay surfaces also depends greatly on soil moisture. High soil moisture, which results in an expansion of clay minerals and provides rather thick layers of adsorption water, facilitates diffusion and the release of interlayer $K^+$ and/or $NH_4^+$.

## B. Biological Factors and Processes

Formula [5] describes the nutrient uptake during a certain time for one unit root length, e.g., 1 m. Total nutrient uptake ($U_t$) of an individual plant for a given time may thus be described by the equation

$$U_t = 2\pi r \cdot \bar{C} \cdot t \cdot l, \tag{7}$$

where $l$ denotes the root length. Hence, the root length appears as an important factor. This actually has been shown by Silberbush and Barber (1983b, 1983c) for phosphate and $K^+$. Based on $K^+$ uptake experiments with soya bean (*Glycine soja*) (Silberbush and Barber, 1983a), these authors developed a $K^+$ and phosphate uptake equation (Barber and

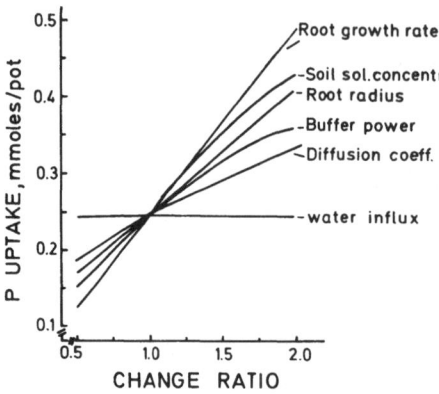

**Figure 4.** Importance of a change in availability parameters on phosphate uptake. Modified after Silberbush and Barber (1983a), *Plant and Soil* 74:93–100. Copyright © 1983 by Martinus/Nijhoff/Dr. W. Junk Publishers. Used with permission.

Cushman, 1981) in which 11 parameters were incorporated. Applying relative rates of these parameters according to 0.5, 1.0, 1.5, and 2.0 and calculating then the uptake of $K^+$ or phosphate, respectively, yielded the results shown in Figure 4. The most important parameters for P uptake corresponded to the following sequence: root length > P concentration in the bulk soil solution > root radius > P buffer power > diffusion coefficient. This emphasizes the important role of root length for the exploitation of phosphate, a parameter that has often been neglected. The finding is in good agreement with recent results of Steffens (1984) who found a highly significant correlation between root length and the P uptake of ryegrass (*Lolium perenne*) ($r = 0.87^{+++}$) and of red clover (*Trifolium pratense*) ($r = 0.91^{+++}$). For $K^+$ uptake a similar picture was obtained. Root length was the most efficient parameter, followed by root radius (Silberbush and Barber, 1983c).

Results and considerations of these authors revealed an additional important aspect of nutrient availability. Taking a root segment as a cylinder, it represents a sink to which nutrients diffuse in a centripetal way. This centripetal direction of nutrient flux results in the increase of nutrient concentration at the root surface as the root radius decreases. Consequently, ion depletion at the root surface will diminish with a decrease in root radius. This actually was shown by Silberbush and Barber (1983c); highest phosphate depletion occurred with corn roots, root $\phi = 200$ μm; followed by grass roots, root $\phi = 50$ μm; followed by root hairs, $\phi = 10$ μm; and finally mycorrhiza hypha, $\phi = 3.3$ μm. At the latter hardly any phosphate depletion was observed. It thus appears that thin roots are much more efficient in $K^+$ and phosphate exploitation than thick ones. Root radius is not only a question of plant species but also depends on soil structure (Peterson and Barber, 1981). Roots growing in compacted soils are often thinner.

Root hairs are also considered very important for mining nutrients from the soil (Nye, 1966; Lewis and Quirk, 1967). As shown in Figure 2 the zone around the root being penetrated by root hairs is depleted of phosphate (Hendriks *et al.*, 1981). Thus, root hairs contribute to the soil volume exploited by the plant. According to Hendriks *et al.* (1981) the depletion zone of corn roots having a mean root hair length of 0.7 mm extended 1.6 mm from the root surface into the soil, while rape (*Brassica napus*) with a root hair length of 1.3 mm had a depletion zone of 2.6 mm. Nevertheless, the importance of root hairs for the acquisition of nutrients may be doubted. In a recent experiment Itoh and Barber (1983) showed that P uptake of carrots (*Daucus carota*), lettuce (*Lactuca sativa*), and wheat (*Triticum aestivum*) agreed satisfactorily with the nutrient uptake model of Barber and Cushman (1981) without considering root hairs. Only in species with a high ratio of "root hair surface/root surface," such as tomato (*Lycopersicon esculentum*) and Russian thistle (*Salsola kali*), did P uptake not correspond with the calculated uptake. However, when the root hair surface in these species was also considered, the calculated P uptake was in good agreement with the measured one. The authors conclude from this finding that only in cases where the root hair surface contributes significantly to the total root surface will root hairs play a major role in the exploitation of soil phosphate. The authors suggest that with wheat, root hairs often extend into soil zones that are already depleted of phosphate. This observation is in good agreement with results of Bole (1973) who found that root hair density had no major impact on P uptake of wheat.

The transition zone between root and soil, called the rhizosphere by Hiltner (1904), differs considerably from the bulk soil. Physico-chemical and biological processes going on in this zone are of relevance for uptake and dynamics of plant nutrients. Exudation of inorganic and organic material by roots into their environment bring about particular conditions for the turnover of plant nutrients. From the release of inorganic material, $H^+$ deserves the paramount interest, since pH changes in the rhizosphere may affect the solubility of plant nutrients as well as multiplication and activity of microbes living at or adjacent to the root surface. From the works of Dijkshoorn (1957), Coic *et al.* (1962), and Kirkby and Mengel (1967), it has been concluded that $H^+$ or $HCO_3^-$ release from roots is primarily a question of cation–anion balance. The theoretical concepts concerning cation–anion balance are well supported by investigations in laboratory as well as in the field. Thus, Smiley's (1974) investigations showed that nitrate nutrition resulted in a pH increase, whereas $NH_4^+$ nutrition decreased the pH in the rhizosphere. The $H^+$ release measured was in the range of $NH_4^+$–N uptake (Dejaegere and Neirinckx, 1978), while the release of $OH^-$ or $HCO_3^-$, respectively, is much lower, for corn about 1/10 of the $NO_3^-$ uptake (Mengel *et al.*, 1983). The pH decrease in the rhizosphere as affected by $NH_4^+$–N nutrition may contribute to the solubilization of phosphates (Riley and Barber, 1971). Symbiontically living legumes take up relatively high amounts of cations while the anion uptake is low owing to the fact that $NO_3^-$ uptake is virtually nil. It is

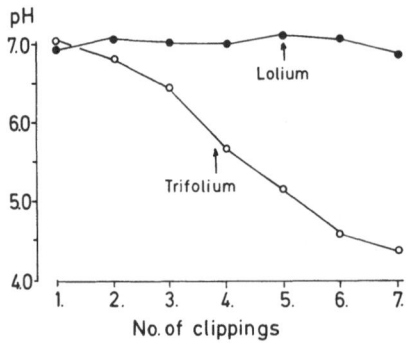

**Figure 5.** Effect of *Lolium perenne* and *Trifolium pratense* on soil pH during a period of seven clippings. After Mengel and Steffens (1982), *Z. Pflanzenern. Bodenk.* 145:229–236. Copyright © 1982 by Verlag Chemie GmbH. Used with permission.

assumed that this excess of cations results in a remarkable release of $H^+$. Actually this has been found by various authors (Israel and Jackson, 1978; Nyatsanga and Pierre, 1973; Andrew and Johnson, 1976; Aguilars and van Diest, 1981). Mengel and Steffens (1982) in studying the effect of ryegrass and red clover on soil pH (Udalf derived from loess) found a considerable pH decrease under red clover during a relative short time, while under ryegrass the pH remained constant (see Figure 5). Ryegrass had been fertilized with $NH_4NO_3$ and red clover had been supplied with symbiontically fixed N. The high cation excess in red clover was still lower than the release of $H^+$ into the soil medium, suggesting that the $H^+$ release was not a simple exchange for absorbed cations. In a more detailed study, Hauter (1983) found that the secretion of $H^+$ into the soil by roots of intact red clover plants was significantly higher than the excess of cations taken up (see Table 1). This may suggest that $H^+$ secretion is not controlled by cation–anion uptake but seems to be an active process. The assumption is supported by results of Mengel and Malissiovas (1982) who found that the $H^+$ release of roots of intact grape vine plants (*Vitis vinifera*) into distilled

**Table 1.** Cation Excess in the Upper Plant Parts and $H^+$ Secretion by Roots of Red Clover Grown Symbiontically on a Luvisol in Pots[a]

| No. of clippings | Soil pH | Cation excess (in mmol pot$^-$) | $H^+$ secretion (in mmol pot$^-$) |
|:---:|:---:|:---:|:---:|
| 1 | 7.0 | 54.9 | 165 |
| 2 | 6.4 | 60.4 | 120 |
| 3 | 5.9 | 43.9 | 70 |
| 4 | 6.0 | 43.7 | – |
| 5 | 5.3 | 19.8 | 70 |
| 6 | 5.2 | 8.1 | 10 |
| Total | | 230.8 | 435 |

[a]The data relate to six clippings obtained during a growing season and to plants per pot containing 7 kg soil. After Hauter (1983).

water was higher in light than in dark. Schubert[1] found that the $H^+$ release of roots is an active process driven by an ATPase. According to these results, plant roots are able to secrete $H^+$ without a concomitant cation uptake, and thus they may influence the direct root vicinity to a remarkable extent. The actual drop in pH at the root surface may be considerable. Hauter (1983), in measuring the pH at the root surface of red clover growing in a medium buffered soil (Udalf), found pH levels about 1 to 2 units lower than in the bulk soil. In a well-buffered Entisol (Fluvent) with free carbonate, however, no pH difference between the root surface of red clover and the bulk soil was found. This finding may show that the actual pH prevailing in the rhizosphere depends much on the $H^+$ buffer power of soils.

Organic matter released by roots can be divided into sugars, amino acids, organic anions, slimes, and the sloughage of root material released when roots grow into the soil (Hale and Moore, 1979). Root sloughage is considered to be the main source of organic C released. According to investigations of Sauerbeck and Johnen (1976) with wheat and mustard (*Sinapis alba*), about 50% of the organic C translocated from the tops into the roots was delivered into the soil in the form of organic C. The organic matter serves as food for microbes, fungi as well as bacteria, and it is for this reason that the microbial density in the rhizosphere is generally higher than in the bulk soil (Rovira and Davey, 1974). From the numerous microbial species living in the rhizosphere, species that are involved in plant nutrient turnover are of particular interest, such as the ammonifying microbes producing $NH_4^+$ from organic N, $N_2$ reducing bacteria (see also p. 80), and also denitrifying bacteria. Some species, e.g., *Spirillum lipoferum* (Neyra and Döbereiner, 1977), are capable of $N_2$ fixation as well as denitrification, depending on the partial pressure of $O_2$ in the rhizosphere. Reducing conditions (anaerobis) that may occur also in upland soils after heavy rainfall may transform the rhizosphere into a zone with high denitrification owing to the fact that in the rhizosphere the denitrifying bacteria will find plenty of digestable organic C (Rolston *et al.*, 1976; Webster and Dowdell, 1982). The risk of N losses by denitrification is especially high in flooded soils. Roots of intact rice plants secrete $O_2$ into the rhizosphere and thus are capable of leaving the root adjacent zone in an oxidized status. This capability of rice roots to prevent too low a redox potential in the rhizosphere is affected by a poor $K^+$ nutrition of rice (Trolldenier, 1973). Thus, under low K supply higher N losses by denitrification may occur.

Recent investigations of Helal and Sauerbeck (1984) have shown that the rhizosphere is relatively rich in organic phosphates, especially inositol phosphates, and that these organic phosphates have a high turnover. This finding is in good agreement with the work of Blair and Boland (1978),

---

[1]Unpublished data.

which showed that mineralization of organic phosphate was favored by the presence of plant roots. Phosphate mobility in the rhizosphere is affected by $H^+$ or $HCO_3^-$ release of roots. A pH decrease may result in a better solubilization of Ca phosphates (Riley and Barber, 1971; Aguilars and van Diest, 1981), but can also have a reverse effect, since low pH promotes phosphate adsorption. Moghimi *et al.* (1978) found 2 oxo-gluconate in the rhizosphere of wheat roots and discovered that it was capable of solubilizing phosphate from hydroxyapatite. Berthelin and Leyval (1982) reported that the microflora of corn roots was able to bring about a $K^+$ release from biotite.

Besides the sloughage, mucilage produced by root tips (Paul and Jones, 1976) contributes much to the total organic C released into the soil. The slimy material may serve as food for bacteria and is, according to Nishizawa *et al.* (1983), of particular importance for the $N_2$-fixing bacteria. Guckert *et al.* (1975) showed by means of an electron micrograph that the bacteria are embedded in the slime and may dissolve it by lytic enzymes. A further important function of the mucilage is that it brings about close contact between roots and soil particles and thus favors water and nutrient transport towards the root. Mycorrhiza is of special importance for the acquisition of water and phosphate. The network of hyphae extending from the root into the soil enlarges the contact area between the soil and the fungus host root association and hence facilitates the acquisition of phosphate and its transport towards the roots (Sanders and Tinker, 1973). Mycorrhizal fungi appear to have little ability to utilize insoluble phosphates such as apatites. They may, however, bring about a release of $K^+$ from biotites as has been shown by Mojallali and Weed (1978) with *Glomus* living at the roots of soybeans. Mycorrhizal growth depends largely on sugars and amino acids released by plant roots. Graham *et al.* (1981) provided evidence that roots of sorghum (*Sorghum vulgare*) released particularly high rates of sugars and amino acids when the plant was poorly supplied with phosphate. The authors showed that an insufficient phosphate supply resulted in an increased permeability of root membranes facilitating the efflux of organic molecules. It is for this reason that soils rich in available phosphate hamper the growth of mycorrhizal fungi.

## III. Nitrogen

### A. Nitrogen Cycle in Nature

Nitrogen is widely distributed in nature. The largest N reservoir represents the atmosphere. As can be seen from Table 2 the N captured in biomass represents only a small proportion of the total N in nature (Delwiche, 1983). Nevertheless, this biomass N is of vital importance. It may exchange

**Table 2.** Distribution of Nitrogen on the Various Spheres[a]

|  | N in gigagram atoms |
|---|---|
| Atmosphere, $N_2$ | $2.8 \times 10^8$ |
| $N_2O$ | 130 |
| Biosphere, plants (land) | 570 |
| plants (ocean) | 14 |
| animals (land) | 15 |
| animals (ocean) | 14 |
| Hydrosphere, organic | $2.4 \times 10^4$ |
| inorganic | $7.1 \times 10^3$ |
| Lithosphere, organic | $5.7 \times 10^7$ |
| inorganic | $1.4 \times 10^7$ |
| Soil, organic | $1.25 \times 10^4$ |
| inorganic | $1.50 \times 10^4$ |

[a]After Delwiche (1983).

quickly with the N of the atmosphere, hydrosphere, and lithosphere. For consideration here, the exchange between the N in the soil profile, the biomass, and the atmosphere are of primary relevance.

## B. Biological $N_2$ Assimilation

Of the large number of living species, only a few are capable of $N_2$ fixation (= $N_2$ assimilation). All are prokaryonts. Of the 47 bacteria families, 11 have species capable of $N_2$ assimilation; and of the 8 families of cyanobacteria, 6 comprise $N_2$-fixing species (Werner, 1980). Most of the $N_2$-fixing organisms are non-symbiontic microbes, and only a few live in symbiosis with higher plants. The most important of the latter are *Rhizobium* and *Actinomyces* species.

Some of the so-called free-living $N_2$-fixing bacteria are strictly anaerobic while others are aerobic. All, except the C autotrophic bacteria, require organic carbon, mainly sugars, polysaccharides, and fatty acids. Carbon supply is frequently the limiting factor for their growth and multiplication. Probably it is for this reason that some of the $N_2$-fixing bacteria live on the surface of plant roots where they profit from the excretion of easily digestable organic C. This relationship between higher plants and bacteria, called association, is not yet completely understood, but it is likely to be very specific. Neyra and Döbereiner (1977) reported that from 33 ecotypes of the tropical gramineae *Paspalum notatum* only 5 stimulated the growth of the $N_2$-fixing bacteria *Azotobacter paspali*. The combination with the suitable ecotype proved to be very efficient, yielding assimilation rates of 60 to 90 kg N $ha^{-1}$ $yr^{-1}$ (Döbereiner *et al.*, 1972).

**Table 3.** Important Free-Living $N_2$-Fixing Bacteria and Their Environmental Requirements

| Genus | pH | $O_2$ | Favorite association or habitat |
|---|---|---|---|
| *Azotobacter* | Neutral | Aerobic | Temperate and tropical |
| *Beijerinckia* | Acid tolerant | Aerobic | Tropical |
| *Azospirillum* | Neutral | Aerobic | Tropical |
| *Bacillus* | Neutral | Aerobic | Temperate |
| *Enterobacter* | Neutral | Fac. anaerobic | Flooded soils |
| *Clostridium* | Acid tolerant | Anaerobic | Flooded soils |
| *Desulfovibrio* | Neutral | Anaerobic | Flooded soils |
| *Anabaena* | Neutral | Aerobic | C autotrophic |
| *Nostoc* | Neutral | Aerobic | C autotrophic |
| *Rivularia* | Neutral | Aerobic | C autotrophic |
| Genus of photo-synthetic bacteria | Neutral | Anaerobic | C autotrophic |

In Table 3 important non-symbiontic $N_2$-fixing bacteria are listed. Most of them are sensitive to low soil pH (pH < 6) and require relatively high temperature for optimum growth. Thus, *Spirillum lipoferum*, living in association with maize (*Zea mays*), has its temperature optimum at 33°C. At 17°C its nitrogenase activity is much depressed and at 10°C completely inhibited (Neyra and Döbereiner, 1977). Consequently, in temperate areas temperature may be a limiting factor in microbial $N_2$ assimilation.

The basic chemical process in $N_2$ fixation is the reduction of $N_2$ to $NH_3$ brought about by the nitrogenase system. This system can be inhibited by $O_2$, and thus higher $O_2$ pressure may depress or even completely block nitrogenase activity. The optimum $O_2$ pressure for $N_2$ assimilation is much below the partial pressure prevailing in the atmosphere. It is for this reason that many $N_2$ species find optimum $O_2$ conditions in flooded soils (Buresh *et al.*, 1980). In the transition zone from the aerobic root surface of flooded rice to the anaerobic bulk soil, $O_2$ partial pressures may prevail that provide favorable conditions for aerobic $N_2$-fixing bacteria such as *Beijerinckia*, *Enterobacter*, and *Azotobacter* species (Savant and DeDatta, 1982). In the anaerobic phase of paddy soils, *Clostridium* and *Desulfovibrio* species as well as C autotrophic bacteria can fix $N_2$, and in the surface water cyanobacteria may bring about $N_2$ fixation. The most important $N_2$-fixing cyanobacteria belong to the genera *Anabaena*, *Nostoc*, and *Rivularia*. According to Watanabe *et al.* (1977), these cyanobacteria play the greatest role in $N_2$ fixation in tropical paddy soils. In Japanese paddy soils, the aerobic $N_2$-fixing bacteria and not the cyanobacteria are believed to contribute most to biological $N_2$ assimilation (Wada *et al.*, 1978). Average

**Table 4.** Nitrogen Fixation Rates in Various Ecosystem, Range in Reported Rates[a]

|  | kg N ha$^{-1}$ yr$^{-1}$ |
| --- | --- |
| Arable land | 7–28 |
| Pasture (non-legume) | 7–114 |
| Pasture (grass–legume) | 73–865 |
| Forest | 58–594 |
| Paddy | 13–99 |
| Waters | 70–250 |

[a]After Hauck (1971), in *Nitrogen-15 in Soil-Plant Studies*, pp. 65–80. Copyright © 1971 by the International Atomic Energy Agency. Used with permission.

rates of biological $N_2$ fixation in flooded soils (paddy) are in a range of 50 kg N ha$^{-1}$ yr$^{-1}$ and thus are much higher than the rates in arable soils that amount to about 12 kg N ha$^{-1}$ yr$^{-1}$ (see also Table 4).

In the transfer of N from the atmosphere to the biosphere, the symbiotic living bacteria, mainly *Rhizobium* and *Actinomyces* species, are of particular importance. Today about 12,000 different leguminous species are known, from which about 200 are agricultural or horticultural crops (Stewart, 1967). This may show the large potential available for the amelioration of known leguminous crops and for the development of new crops. The efficiency in $N_2$ fixation of symbiotic living bacteria depends much on the suitable strain. In the process of $N_2$ reduction by nitrogenase, some $H^+$ is also reduced to $H_2$. The capability of recycling this $H_2$ by hydrogenase and using the $e^-$ produced for the reduction of $N_2$ is essential for the efficiency of $N_2$ reduction (Schubert *et al.*, 1978). *Rhizobium* living in the slime layer of leguminous roots may infect the root tips. The infection is a critical process, and from the large number of *Rhizobium* bacteria present only a few succeed in bringing about an infection (Bauer, 1981). It depends on soil pH, the presence of mineral N in the soil, especially $NO_3^-$, and on $Ca^{2+}$. The latter favors the infection, while inorganic N has the reverse effect (Russell and Johnson, 1975). According to Munns (1968) $NO_3^-$ impedes the development of the infection thread.

The symbiosis of *Rhizobium* or *Actinomyces* species with higher plants suggests some major advantages. First, there is a permanent supply of digestible organic C. Then, there exists a controlled $O_2$ supply brought about by leghämoglobin (Parthier, 1978). Finally, the produced $NH_3$ is rapidly translocated out of the bacteroid so that no major accumulation of $NH_3$, glutamate, or glutamine occurs (Antoniw and Sprent, 1978). This is of importance, since these compounds block the Nif genes responsible for the synthesis of nitrogenase. Likely it is for this reason that the efficiency of $N_2$ fixation is much higher in the symbiotic-living than in the free-living bacteria. The rate of $N_2$ fixation depends also on the nutritional status of the higher plant. Plants with a high photosynthetic activity also translocate the photosynthate into the roots at high rates, thus supplying the *Rhizobium* bacteroids with plenty of energy, which has a favorable impact on $N_2$ assimilation (Feigenbaum and Mengel, 1979).

In Table 4 the $N_2$-fixing potential of various ecosystems is shown (Hauck, 1971). Highest $N_2$-fixation rates were found in leguminous pastures. However, considerable amounts of $N_2$ can be assimilated in forests with leguminous species (*Robinia, Acacia*) or alder (*Alnus*). Daly (1966) reported that under favorable conditions *Actinomyces alni* living in symbiosis with alder (*Alnus rugosa*) fixed about 150 kg N ha$^{-1}$ yr$^{-1}$. According to Werner (1980), roots of angiosperms extending into marine sediments may attain fixation rates as high as 700 kg N ha$^{-1}$ yr$^{-1}$. The cyanobacterium *Anabaena azollae* living in symbiosis with *Azolla pinniata* may assimilate as much as 800 kg N ha$^{-1}$ yr$^{-1}$ under favorable conditions (Singh, 1979). For optimum $N_2$ fixation, an abundant phosphorus supply for the fern is needed (Watanabe *et al.*, 1980). This unique symbiosis is of major importance for lowland rice (Moore, 1969).

### C. Organic Soil N and Mineralization

Organic N represents the largest fraction of soil N. Its content may vary in a wide range of 0.2 to 5 g N kg$^{-1}$ soil. Two major factors control the organic N content of soils: climate, especially temperature; and soil cultivation. Generally soil organic N decreases with an increase of the average yearly temperature. Intensive cultivation and the growth of annual crops with a relatively poor root mass (potatoes, beets) result in a decrease of organic N, whereas perennial crops such as grasses and legumes have a favorable effect on soil organic N. Weller (1983) reported that the mulching of grass in an orchard led to a significant increase of organic N in the upper soil layer (0 to 5 cm) from 2 g N kg$^{-1}$ soil to 3.9 g N kg$^{-1}$ soil after an experimental period of 8 yrs. In some cases, contents as high as 5.3 kg N kg$^{-1}$ soil were attained, whereas in the soil with no grass cover soil organic N amounted only to 2.7 g kg$^{-1}$. These experiments also revealed that the content of organic N in the upper soil layer is not so much a question of mineral N fertilizer application as of keeping the soil free from any vegetative cover or of growing a grass cover that may be mulched or removed. Even in the treatment without mineral N fertilizer, the organic soil N was substantially raised during a period of 8 yrs under a mulched grass cover. It is suggested that under the conditions of grass mulching $N_2$-fixing bacteria may find favorable conditions. In the Rothamsted long field experiments, a clear increase of organic soil N was found on sites that remained undisturbed for about 100 yrs. Some important data from these field trials are presented in Table 5, showing that at Geescroft the increase in soil organic N was less than at Broadbalk. The latter soil contains $CaCO_3$, which probably provides favorable pH conditions for biological $N_2$

**Table 5.** Change in Soil Organic N and pH of Two Undisturbed Sites during an Experimental Period of about 80 Yrs[a]

| Site 1 year | g N kg$^{-1}$ soil | N gain, kg N ha$^{-1}$ yr$^{-1}$ | pH |
|---|---|---|---|
| Broadbalk | | | |
| 1881 | 1.07 | 52 | |
| 1904 | 1.42 | 52 | |
| 1964 | | | |
| Wood | 2.54 | 33 | |
| Meadow | 2.60 | 34 | |
| | | | |
| Geesecroft | | | |
| 1883 | 1.16 | 21 | 7.0 |
| 1904 | 1.31 | 21 | 6.1 |
| 1964 | 1.66 | 15 | 4.5 |

[a]Since 1904 the Broadbalk site has been divided into "wood" and "meadow." From the latter shrubs and trees were regularly removed. Modified after Russell (1973). Reproduced from *Soil Conditions and Plant Growth*, 10th ed., by permission of Longman.

fixation. At the Geescroft location a substantial decrease in soil pH occurred during a period of 80 yrs. The low N accumulation rates (15 kg N ha$^{-1}$ yr$^{-1}$) found during the last decades may be due to the low soil pH.

Organic soil N is a very heterogeneous matter of which the main components are humic acids (about 40 g N kg$^{-1}$), fulvic acids (about 7 g N kg$^{-1}$), non-humic material with widely varying N contents, and the biomass with a N content of about 50 to 100 mg N kg$^{-1}$ dry matter. These organic N-containing compounds are prone to mineralization to differing extents. Fulvic and humic acids are considered to be very stable while the biomass has a quick turnover. Beck (1983) assumes that all N being mineralized results from the biomass fraction. According to Campbell (1978), it is mainly the soil N fraction hydrolyzable by acid hydrolysis that is attacked by soil organisms. This fraction is about 16% of the total organic soil N. It represents the potential of mineralizable N and is much higher than the annual net release of inorganic N resulting from mineralization in cultivated soils. Also, the various incubation methods for estimating the available N in the soil, such as the anaerobic incubation technique of Waring and Bremner (1964) and Ellenberg's (1964) incubation test, which is carried out under field conditions, provide N mineralization rates much higher than those occuring under field conditions. Stadelmann *et al.* (1983) found in normal cultivated soils, tested according to the technique of Waring and Bremner (1964), N mineralization rates of 142 to 814 kg N ha$^{-1}$ representing 1.2 to 7.4% of the total organic soil N. The mineralization rate was significantly correlated with the total organic soil N ($r^2 = 0.821$).

Net mineralization depends on various factors. Generally, soils from permanent grassland show higher mineralization rates than arable soils. Beck (1983) found a significant correlation between the mineralizable N according to the technique of Keeney and Bremner (1966) and the biomass ($r = 0.96$) and the protease activity ($r = 0.98$) in the soil.

It should be kept in mind that the net mineralization of N results from the difference between the decomposition of organic N to inorganic N and the so-called immobilization of inorganic N. The latter can be a biotic process, e.g., the assimilation of $NH_4^+$ or $NO_3^-$ by soil microorganisms; immobilization may also occur by abiotic processes, e.g., by binding inorganic N to organic soil matter. Bremner and Nelson (1968) found in incubation experiments that $NO_2^--N$ was incorporated by lignin. Similar results were reported by Führ and Bremner (1964). In more recent experiments Haider and Farooq-e-Azam (1983) reported that glucose and proteins added to the incubated soil resulted in a quick immobilization of $NH_4-N$. The immobilized N, however, was to a large degree (about 60%) remineralized during the following 24 weeks. Addition of straw and lignin to the incubation media also led to a significant immobilization of $NH_4-N$. In contrast, however, to glucose, the rate of remineralization of the immobilized N was low in the treatment with straw or lignin additions and amounted to only about 20% of the $NH_4-N$ added to the incubation media. The lignin, which was [14]C-labeled in the ring, was hardly decomposed, as was evidenced by the release of [14]$CO_2$ that amounted only to 0.1% of the total C label added. Haider and Farooq-e-Azam (1983) concluded that lignin is hardly used for the synthesis of biomass, but that it contributes substantially to the formation of humus. One can speculate from these results that the N bound by lignin is also used for the synthesis of humic acids and thus hardly available to plants.

This aspect is of particular relevance for the incorporation of straw into soils, today a widely used practice in many countries. It is well known that straw incorporation leads to an immobilization of inorganic N in the soil (Myers and Paul, 1971). The general belief is that the immobilized N will be remineralized earlier or later and contribute to crop nutrition. This thesis, however, has not been proved. At least a portion of the immobilized N will be bound very strongly and thus probably will not be readily available to plants. This has been recently found in incubation and pot experiments by Scherer and Mengel (1983), in which nitrate fertilizer was labeled with [15]N. After an incubation period of 10 months, the highest N label was found in the unidentified fraction of the hydrolyzable N in the treatment with straw application. In the treatment without straw the highest label was in the amide-N fraction. It thus appeared that straw application led to a remarkable N binding by a fraction that mainly contains fulvic and humic acids, while in the treatment without straw the amide-N fraction was enriched with labeled N. This amide fraction is supposed to belong to the biomass. The soil fertilized with labeled $NO_3^-$ and treated with or without straw and incubated for 10 months was then intensively cropped by two clippings of *Lolium perenne* followed by *Sinapis alba*. These crops received only small amounts of inorganic N in order to obtain a high uptake of soil-borne N. Plant growth in the treatment with straw application was much poorer than in the treatment without straw, and so was the recovery of labeled fertilizer N. Recovery of N in the treatment with straw was 38.2%; compared with 75.2% in the treatment without straw. Such stable organic N compounds may even be present in slurries, as was recently revealed by Amberger *et al.* (1982). Incubation studies of Tomar and Soper (1981a) also provided evidence that incorporation of organic matter decreased the availability of urea N for barley (*Hordeum vulgare*). The authors suggest that a close contact between the organic matter and fertilizer N results in N immobilization. This assumption was supported by field experiments of Tomar and Soper (1981b), in which the recovery of urea N was especially low in the treatment in which urea was surface broadcast and the straw incorporated into the soil. Straw application decreased the availability of fertilizer N as well as of soil N.

Whether straw application results in the formation of very stable organic N compounds may also depend on the biological activity of soils. Graff and Kühn (1977) reported that earthworms diminished the immobilization of inorganic N brought about by straw. Research data of long-term field trials with straw incorporation into the soil showing that straw depressed yields on soils with a low biological activity (Bachthaler and Wagner, 1973) are in good agreement with the assumption that inorganic N is fixed by lignin, particularly under conditions of a poor biological activity in the soil.

The process may lead to a diminished N availability in the soil. On the other hand, it may improve soil structure. As evidenced by Biederbeck *et al.* (1980) in a 20-yr field experiment, the plots that had received straw showed an improved soil structure while the soils of plots without straw had a higher content of exchangeable $NH_4^+$.

Mineralization of organic N starts with ammonification and under aerobic conditions and a favorable soil pH is followed by nitrification. From a microbiological view the processes of ammonification and nitrification differ considerably. Ammonification is carried out by a broad spectrum of different C heterotrophic organisms; nitrification only by a small number of C autotrophic bacteria. Temperature has a decisive influence on nitrification and ammonification, the latter having its maximum at about 50°C, whereas the temperature maximum for nitrification is at 26°C (Beck, 1983). This has an impact on $NO_3^-$ and $NH_4^+$ accumulation in the soil, as can be seen from Figure 6. In aerobic soils of temperate areas hardly any $NH_4^+$ accumulates in the soil because the $NH_4$–N produced is quickly oxidized to $NO_3^-$. Under such conditions $NH_4$–N production is the rate-limiting step in nitrification. In soils, however, with temperatures $> 30°C$, ammonification is faster than nitrification and here even under aerobic conditions $NH_4$–N may accumulate. The high ammonification rate at high temperature is indicative of the rapid decomposition of organic matter in tropical soils.

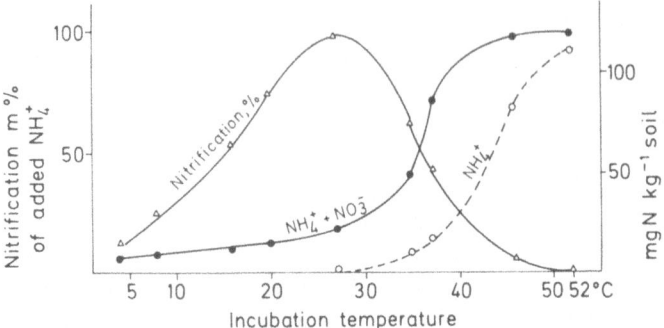

**Figure 6.** Effect of temperature on nitrification in percent of added $NH_4^+$, on the accumulation of $NH_4^+ + NO_3^-$, and on the accumulation of $NH_4^+$ in the soil during an incubation period of 21 days. After Beck (1983), *Z. Pflanzenern. Bodenk.* 146:243–252. Copyright © 1983 by Verlag Chemie GmbH. Used with permission.

### D. Fixation and Release of $NH_4^+$ by Clay Minerals

Clay minerals, particularly illites and vermiculites, bind $NH_4^+$ very strongly and selectively in the interlayer between the tetrahedral silicate layers of 2:1 clay minerals (Stanford and Pierre, 1946; Bremner, 1959; Rodrigues, 1954). According to various authors, more N exists in many soils as clay mineral fixed $NH_4^+$ than all other forms of soil N (Hinman, 1964; Dalal, 1977a). The so bound $NH_4^+$ is called "fixed $NH_4^+$," although this term is not correct since it suggests a non-available N fraction that, as has been shown by recent experiments, is not the case. It is more appropriate to term this fraction "non-exchangeable $NH_4^+$" (Mengel and Scherer, 1981) or "intercalary $NH_4^+$" (Osborne, 1976). This $NH_4^+$ fraction is not exchangeable by $K^+$. Nevertheless, $NH_4^+$ can be slowly released from this fraction under partciular conditions in an analogous way as $K^+$ is released from the interlayers of 2:1 clay minerals (Scott and Smith, 1966). Allison *et al.* (1953) and also Schwertmann (1966) reported that a small fraction of the non-exchangeable $NH_4^+$ is accessible to microbial oxidation. According to Guo *et al.* (1983), the rate of nitrification of non-exchangeable $NH_4^+$ is much slower than that of the exchangeable $NH_4^+$ fraction. Legg and Allison (1959) found that the non-exchangeable $NH_4^+$ plays no major role in plant nutrition. This observation contrasts with field experimental data of Kowalenko and Cameron (1978), who found that a substantial amount of fertilizer $NH_4^+$ was bound in a non-exchangeable (fixed) form and later made available to growing barley. Van Praag *et al.* (1980) also reported that a luvisol derived from loess released remarkable quantities of non-exchangeable $NH_4^+$ during a growth season. Some of the most interesting data obtained by van Praag *et al.* (1980) are shown in Table 6. There was a

**Table 6.** Change in the Content of Non-Exchangeable $NH_4^+$ (Fixed $NH_4^+$) during a Cropping Period in the Soil Profile[a]

| Profile depth (m) | Under wheat harvest, 1972 | Under oats harvest, 1973 |
|---|---|---|
| | ($mg\ N\ Kg^{-1}$ dry soil) | |
| 0–0.1 | 80.5 | 81.4 |
| 0.1–0.2 | 89.3 | 81.4 |
| 0.2–0.3 | 92.8 | 69.1 |
| 0.3–0.4 | 104.0 | 62.5 |
| 0.4–0.5 | 110.0 | 63.9 |
| 0.5–0.6 | 115.0 | 88.1 |
| 0.6–0.7 | 127.8 | 98.0 |

[a]Modified after van Praag *et al.* (1980), *Soil Science* 130:100–105. Copyright © 1980 by The Williams & Wilkins Co., Baltimore. Used with permission.

remarkable decrease in non-exchangeable $NH_4^+$ from one year to the other, particularly in the deeper soil layers. The total quantity of $NH_4^+$ released from the profile during one year amounted to about 250 kg N ha$^{-1}$. It is noteworthy that the net release of $NH_4^+$ was especially high in deeper soil layers. The pattern of $NH_4$ depletion from the profile may suggest that the deeper soil layers in particular may release $NH_4^+$, and the top layer does not. One has to keep in mind, however, that positions of the top layer depleted of $NH_4^+$ may be quickly refilled by $NH_4^+$ originating from mineralization. This is likely to occur as can be seen from Figure 7 (Mengel and Scherer, 1981). The non-exchangeable $NH_4^+$ was determined in an alluvial soil during a growth period at certain intervals. In the two upper layers (0 to 0.3 m and 0.3 to 0.6 m depth), a highly significant decrease in non-exchangeable $NH_4^+$ occurred from February until May. In the following months a substantial increase of the non-exchangeable $NH_4^+$ was

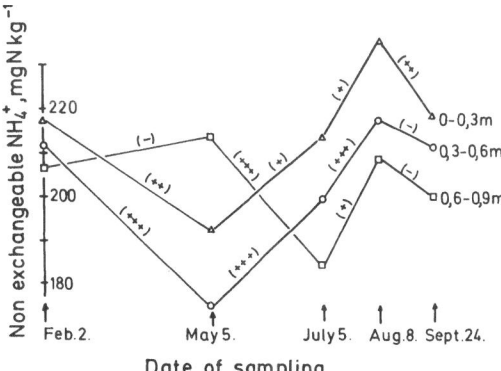

**Figure 7.** Concentration of non-exchangeable $NH_4^+$ during a growing season under oats in three layers of an alluvial soil. After Mengel and Scherer (1981), *Soil Science* 131:226–232. Copyright © 1981 by The Williams & Wilkins Co., Baltimore. Used with permission.

observed, and in August the content of non-exchangeable $NH_4^+$ was as high as at the beginning of the experiment (0.3 to 0.6 m) or even higher (0 to 0.3 m). The soil was rich in organic matter (about 6%) and well known for its high N-mineralization rate. The study found evidence using $^{15}N$-labeled $NH_4^+$, that under field conditions significant quantities of non-exchangeable $NH_4^+$ were released during the first months of the growth period, and later the $NH_4^+$-depleted positions of the clay interlayers were refilled with non-labeled $NH_4^+$. Figure 7 also reveals that the highest depletion in non-exchangeable $NH_4^+$ found in the deepest soil layer (0.6 to 0.9 m) occurred about 2 months later, which is in accordance with the root development of oats grown on this soil.

The depleted clay minerals were not refilled by $NH_4^+$ in all cases during a growth period. This depends greatly on the rate of N mineralization during a season. In a luvisol derived from loess that had received in the preceding autumn a heavy slurry application, Mengel and Scherer (1981) found a significant decline in non-exchangeable $NH_4^+$ under winter barley. Also in this case the depletion from the deeper layer (0.6 to 0.9 m) occurred at a later stage. On this soil a small but significant refilling of clay with $NH_4^+$ was found in the upper soil layer (0 to 0.3 m) but not in the deeper layers (0.3 to 0.9 m) at the end of the growth period. The net release of non-exchangeable $NH_4^+$ amounted to almost 500 kg N $ha^{-1}$ per season.

Studies carried out in flooded rice soils also showed that substantial amounts of non-exchangeable $NH_4^+$ can be released as evidenced with $^{15}N$ provided to crops. Highest rates released, ranging from 80 to 100 kg N $ha^{-1}$ per season, were found in an alfisol with vermiculite as the main clay mineral, whereas two vertisols with smectite as the dominant clay mineral only released 20 to 50 kg N $ha^{-1}$ non-exchangeable $NH_4^+$ during the season (Keerthisinghe et al., in press). Scherer (1980) reported that soils of basaltic origin with smectite as the dominant clay mineral, although not low in nonexchangeable $NH_4^+$, released only minute amounts of interlayer $NH_4^+$. Osborne (1976) also reported that the release of non-exchangeable $NH_4^+$ depended largely on the mineral, with illite and vermiculite being the most important for $NH_4^+$ release.

It is assumed that the process of $NH_4^+$ release from the interlayer is analogous to the process of $K^+$ release, which is an exchange and a diffusion in the interlayer (Scott and Smith, 1966). Probably the edge zones of the clay margins are particularly prone to $NH_4^+$ binding and release. A net release will only occur if the $NH_4^+$ concentration in the adjacent soil solution is low. This generally will occur when $NH_4^+$ is absorbed by plant roots with high rates. Consequently, depletion in non-exchangeable $NH_4^+$ occurs when plants are grown on a soil. It is supposed that particularly in the root vicinity the $NH_4^+$ concentration is decreased to a very low level so that net $NH_4^+$ release may occur. Whether nitrification may also lower the $NH_4^+$ concentration of the soil solution to an extremely low level is doubted. It is likely that the release of non-exchangeable $NH_4^+$ by microbial oxidation of $NH_4^+$ proceeds only at low rates. High soil moisture promotes the release of non-exchangeable $NH_4^+$ (Scherer and Mengel, 1981), whereas high levels of exchangeable soil $K^+$ have the reverse effect.

Release of non-exchangeable $NH_4^+$ and refilling of clay interlayers with $NH_4^+$ have some important agronomic aspects. The $NH_4^+$ produced by mineralization or applied as fertilizer before winter or monsoon rainfall and being subsequently fixed by clay minerals is protected against nitrification, and thus the non-exchangeable $NH_4^+$ will be available to a crop during a growing period.

## E. Nitrogen Losses from the Soil System

Nitrogen losses from the soil may be caused by leaching and by volatilization of various N forms. Nitrate is especially prone to leaching while $NH_4^+$ leaching may occur only on sandy and organic soils since they are less capable of $NH_4^+$ adsorption. The leaching rates depend largely on climate and weather conditions, on the drainage rates of water, and also on the quantity of soluble N, mainly $NO_3^-$, in the top layer of the soil. To avoid major N-leaching losses, N fertilization should be adjusted to the N requirement of the crop. The implementation of such a N fertilizer policy, which may include split application of fertilizer N, is relatively easy with mineral fertilizer but is difficult when farmyard manures or slurries are applied (Sluijmans and Kolenbrander, 1977; Pfaff, 1963). Losses are especially high when these organic fertilizers are applied in late summer or autumn so that some of the organic N can still be converted to $NO_3^-$ and is thus prone to leaching by winter rainfalls. Losses can be diminished by the application of nitrification inhibitors as has been recently suggested by Amberger and Vilsmeier (1979). High nitrogen leaching may also occur under deteriorated leguminous crops or when a leguminous crop is incorporated into the soil (Low and Armitage, 1970). Nitrogen losses resulting from leaching are a particular problem in the tropics where light-textured soils and heavy rainfall are frequent.

Under temperate climate conditions on soils with a medium to high clay content, drainage rates of water even during the winter season are low and so are the amounts of leached nitrogen (Teske and Matzel, 1976). Riga *et al.* (1980), carrying out lysimeter experiments with labeled N over several years under field conditions on a luvisol derived from loess, found that denitrification rather than leaching may lead to major N losses. In this experiment about 20 to 40% of the fertilizer N was lost by denitrification. Rolston *et al.* (1976) found that significant losses of $NO_3^-$ fertilizer may occur under field conditions if favorable conditions for denitrification prevail after fertilizer application. Nitrate application resulted in a denitrification loss of about 50% of the fertilizer nitrogen under moist soil conditions ($-2$ to $-7$ kPa). Ammonium fertilizer labeled with [15]N applied under the same conditions did not result in any measurable release of labeled $N_2O$ or $N_2$ (Rolston, 1977). A similar observation has been reported by Riga et al. (1980) who found under field conditions that the recovery of [15]N-labeled N was higher with $NH_4^+$ than with $NO_3^-$–N fertilizer. Kjellerup and Dam Kofoed (1983) found in lysimeter experiments run for several years under field conditions in Denmark that of the total fertilizer N applied, about 60% was taken up by the crop, 20% was incorporated into the soil, and 5% was leached. The other 15% was lost by denitrification. Denitrification losses depend greatly on the availability of digestable organic carbon. In this respect, straw incorporation into the soil is of particular interest. Ganry et al. (1978) found in lysimeter experiments that the incorporation of pearl millet (*Sorghum vulgare*) straw into the soil led to an increase in denitrification losses under the tropical conditions of the Senegal. Scherer and Mengel (1983) found in pot experiments with labeled $NO_3^-$ fertilizer that incorporation of straw into the soil resulted in a decrease of the $NO_3^-$ level in the soil but did not promote denitrification. Schmeer (1983) studied denitrification under a controlled Ar-atmosphere with soil quantities of about 12 kg soil. He found that denitrification losses owing to straw incorporation occurred only under highly reducing soil conditions that led to the release of $CH_4$. Schmeer (1983) suggested that such anaerobic conditions hardly occur in arable upland soils but may prevail in flooded soils. Thus, Araragi and Tangcham (1979), when investigating gas release under rice in a paddy field, found that in the treatment with straw incorporation, the $N_2$ release by the soil was about 5 to 10 times higher than in the treatment without straw. $NH_4^+$ fertilizer, however, had no major effect on the $N_2$ release. Treatments using straw had a $CH_4$ release that was about 50 to 100 times higher than the treatments without straw. Methane development started when the $N_2$ release slowed down, a finding that may indicate that the $N_2$ release resulted from straw oxidation.

    $NH_4$ in flooded rice soils may be oxidized in the aerobic top layer of the soil and in the aerobic rhizosphere of rice roots (Reddy et al., 1976). The limiting process of denitrification in flooded soils is the diffusion of $NH_4^+$ into the aerobic zones. Reddy and Rao (1983) in studying ammonification and N flux in an organic soil under submerged conditions found a flux rate of 45 mg $NH_4$–N $m^{-2}day^{-1}$. The rate of mineralization of organic N was about twice as high. Under the particular experimental conditions, less than 7% of the mineralized N was recovered as inorganic N in the flood water, and the remaining N was lost from the system. Calculated on a hectare basis, the loss was about 350 kg N $ha^{-1}$ $yr^{-1}$. This may demonstrate the remarkable drain of N from the soil system into the atmosphere in systems rich in organic N under anaerobic conditions.

    Denitrification losses may be high on permanent grassland after $NO_3^-$ application, especially if this is followed by rainfall (Webster and Dowdell, 1982). Rolston et al. (1976) reported that the lowest oxygen concentration under ryegrass grown in the field occurred in the upper soil layer (0 to 0.1 m), with $O_2$ concentrations of 10 to 14% $O_2$, while in the deeper soil layer (0.4 to 1.8 m) the $O_2$ concentration of the soil atmosphere amounted to 18%. Thus, the upper soil layer of grass swards provides not only plenty of organic C but also the reducing conditions necessary to promote denitrification.

The evolution of N gases occurred rather quickly after N application in the field experiment of Rolston *et al.* (1976) as shown in Figure 8. The peak of $N_2O$ release occurred about 8 days after $KNO_3$ application, and $N_2$ peaked after about 18 days. This sequence suggests that $N_2O$ is a precursor of $N_2$. The finding makes evident that high N losses may occur also under practical farming conditions if nitrate fertilizer application is followed by rainfall. Figure 8 also shows that $N_2O$ represents only a small proportion of the gaseous nitrogen released by denitrification; about 5% according to Rolston (1977) and less than 1% according to Schmeer (1983). $NO_2$ is an intermediary product in the microbiological reduction sequence (Gaskell *et al.*, 1981):

$$NO_3^- \rightarrow NO_2^- \rightarrow N_2O \rightarrow N_2.$$

Volatilization losses in the form of $NH_3$ have been comprehensively treated by Terman (1979) in a valuable review paper, so only some major points will be raised here. Losses depend much on soil pH, which controls the equilibrium:

$$NH_4^+ \leftrightarrows NH_3 + H^+.$$

Thus, with an increase in pH, the hazard of N loss by $NH_3$ volatilization increases. Soils with a higher cation exchange capacity generally have relatively low $NH_4^+$ concentrations in the soil solution, and thus $NH_3$ losses are also lower than in sandy soils. Type of N fertilizer and its turnover in the soil have an impact of $NH_3$ losses. According to Terman (1979), urea-N in particular may be easily lost because the hydrolysis of urea may lead to $(NH_4)_2CO_3$, of which both components, $NH_3$ and $CO_2$, are volatile.

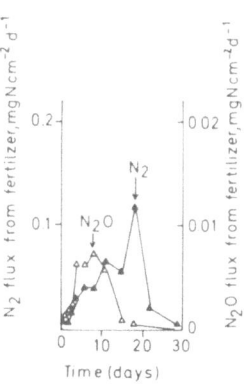

**Figure 8.** Release rates of rates $N_2$ and $N_2O$ related to the time after fertilizer application. After Rolston *et al.*, *Soil Science of America Journal*, Volume 40, 1976, pages 259–266, by permission of the Soil Science Society of America.

Ammonia released by urease may also increase soil pH by protonation. Thus, Overrein and Moe (1967) have shown a pH increase around a urea granule. Such local spots with a high pH level adjacent to a $NH_4^+$ source may lead to considerable N losses, particularly when urea is surface applied. This is the reason why even on soils with a pH below 7, volatilization losses of $NH_3$ after urea dressing are likely. Deep application is recommended so that the $NH_3$ released from the granule may be captured by adsorption sites on its way to the soil surface. Particularly high losses of volatile $NH_3$ may occur when urea is applied together with a nitrification inhibitor as has been recently found by Rodgers (1983) in laboratory experiments. According to Chin and Kroontje (1963) the chemical hydrolysis is proceeding at low rates, so the rates of volatile losses should also be low. Hydrolysis by urease, however, proceeds quickly. Consequently, urease activity may control $NH_3$ volatilization losses. It is for this reason that, with an increase in soil temperature and thus in urease activity, the risk of $NH_3$ losses from urea is also increased. Microbial urease activity depends also on the presence of digestible organic carbon in the soil.

Ammonium fertilizers that form more stable products in the soil are less prone to $NH_3$ volatilization. This is true for $(NH_4)_2SO_4$, $NH_4NO_3$, and especially for $NH_4$-phosphates. The latter may form $Ca(NH_4)_2 \cdot (HPO_4)_2 \cdot H_2O$, a relatively stable compound (Terman, 1979). Ammonium resulting from the decomposition of soil organic N may also be lost by volatilization, particularly under high soil pH conditions. The same is true for organic fertilizers. According to Lauer et al. (1976, quoted in Terman, 1979), farmyard manure may lose about 60 to 90% of its $NH_4$-N when surface applied in summer. Slurries, generally having a pH > 7, easily release a substantial proportion of their $NH_4$–N when applied on sunny or windy days. High $NH_3$ losses may occur in flooded soils, particularly in fields with vigorous algae growth (Savant and DeDatta, 1982). Mikkelsen et al. (1978) found diurnal changes of the pH in flood water of rice. On Maahas clay (Philippines) a maximum pH level as high as 9 was found at noon, while at midnight the minimum pH level was about 7. Such a high pH promotes $NH_3$ volatilization considerably and may lead to substantial losses of fertilizer N. It is assumed that the pH peak at noon is associated with the photosynthetic activity of algae. In cases in which the soil is less favorable to algal growth and where the pH of the flood water is < 7, there is no danger of major $NH_3$ losses (Mikkelsen et al., 1978).

Finally, it should be mentioned that even in the plant $NH_3$ can be produced and released to the atmosphere. According to Stutte et al. (1979), the process is favored by high temperature and transpiration. These authors computed $NH_3$ losses from soybean foliage as high as 45 kg N ha$^{-1}$ per season. Hooker et al. (1980b) found in experiments with winter wheat that especially during the grain filling period $NH_3$ is released into the atmosphere. It is assumed that during plant senescence $NH_3$ is produced by protein decomposition. Ammonia may also be absorbed from the atmosphere. Release from the leaves or absorption by leaves depends on the partial pressure of $NH_3$ in the atmosphere (Farquahr et al., 1980). Release and absorption of $NH_3$ by aerial plant parts may bring about a nitrogen exchange between longer distances. Lemon and van Houtte (1980) suggest that of the total $NH_3$ recycled from the atmosphere into the biosphere or soil, the major proportion originates from the $NH_3$ release of upper plant parts.

## F. Determination of Available Nitrogen

From the relatively large amount of N present in the rooting zone of soils, only a small proportion is directly available to plants. This is $NO_3^-$, mainly present in the soil solution and the dissolved and adsorbed $NH_4^+$. On aerated cultivated soils with not too low a pH, $NH_4$-N is rapidly converted to $NO_3^-$, so in these soils $NO_3^-$ is the most important directly available N form. The nitrate content is affected by various processes (Page and Talibudeen, 1977) such as vertical transport owing to evaporation or leaching, nitrification, denitrification, and uptake and assimilation by higher plants and microbes. On representative arable soils in the temperate zone a typical trend of the nitrate concentration in the soil throughout the growth season has been found. In fallow soils the $NO_3^-$ concentration increases from spring until the end of August. In late summer $NO_3^-$ accumulation in the soil may amount to as much as 150 kg N $ha^{-1}$ in the rooting zone of fertile fallow soils (Winner et al., 1976). In cropped soils generally no $NO_3^-$ accumulation occurs during the growth season because of $NO_3^-$ uptake by the plant. In winter cereals highest $NO_3^-$ concentrations are found in the spring, decreasing with an advance in plant growth, and attaining a minimum with the onset of flowering (Gutser and Teicher, 1976; Aichberger, 1982). At this time the $NO_3^-$ quantity in the soil profile (0 to 0.1 m) is about 20 to 30 kg N $ha^{-1}$. Fertilizer N application is clearly reflected by the $NO_3^-$ level in the soil showing a steep increase after $NO_3^-$ and also $NH_4^+$ application. Often a priming effect has been observed; that means the increase in available N in the soil following a N fertilizer application was higher than the N quantity applied (Broadbent and Nakashima, 1971; Gutser and Teicher, 1976; Aichberger, 1982). Figure 9 shows two typical curves of nitrate concentration during a growing season under sugar beet; one curve representing the fertilizer treatment (160 kg N $ha^{-1}$) and one representing the $N_0$ treatment (Bronner, 1974). Fertilizer application in spring may increase the $NO_3^-$ in the rooting zone to about 200 kg N $ha^{-1}$. But even such high levels are depleted to a low level of about 30 kg N $ha^{-1}$ by a good crop stand during 3 to 4 months of vigorous growth.

Nitrate content in the rooting zones of soils in spring has been used by several authors as a basis for N fertilizer application (Soper and Huang, 1962; Borst and Mulder, 1971; Sims and Jackson, 1971; Scharpf and

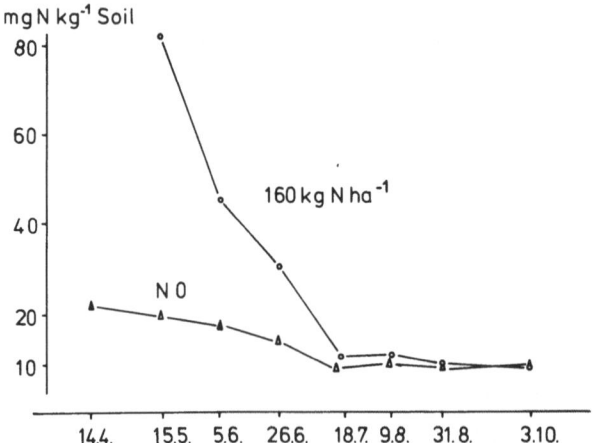

**Figure 9.** Soluble nitrogen, extracted with a NaCl solution, in soil under sugar beet throughout the growing season in a treatment without N fertilizer ($N_0$) and in a treatment with N fertilizer. After Bronner (1974), *Landw. Forsch. SH* 30(II):39–44. Copyright © 1974 by J.D. Sauerländer's Verlag. Used with permission.

Wehrmann, 1975; Müller *et al.*, 1976; Fox and Piekielek, 1978; Wehrmann and Scharpf, 1979; Roberts *et al.*, 1980). The principle of the method ($N_{min}$ method) is based on the fact that the inorganic N ($NO_3^-$ and $NH_4^+$) analyzed in early spring behaves like fertilizer N and can be taken into consideration when computing the N fertilizer rate. If, for example, a winter wheat requires 120 kg N $ha^{-1}$ for its vegetative development (development until anthesis) and 40 kg "$N_{min}$" $ha^{-1}$ have been found in spring by the soil test, the difference of 80 kg N $ha^{-1}$ is recommended for application. The method assumes that no major N quantities are mineralized during the growth period, which actually is the case for cereals for many inorganic soils. Aichberger (1982), in studying the N-balance of winter wheat in a typical luvisol, found that in the treatment with no N fertilizer about 40 to 50 kg N $ha^{-1}$ were released by the soil and taken up by the crop. In the treatment with 80 kg N $ha^{-1}$, fertilizer N and nitrate N in the soil analyzed in spring corresponded with the N uptake of the crop, while in the highest N treatment (160 kg N $ha^{-1}$) the quantity of N taken up by the crop was about 50 kg N $ha^{-1}$ lower than the quantity of fertilizer N applied + nitrate found in the soil in spring. Aichberger (1982) suggests that this deficit was not lost by leaching or denitrification but was fixed biologically.

Soils rich in organic N release considerable amounts of inorganic N during spring. On such soils the $N_{min}$ method is not suited for computing the N fertilizer rate. Soils regularly supplied with farmyard manure also mineralize substantial quantities of organic N during spring and summer. On such soils the $N_{min}$ method may recommend fertilizer rates that are too high (Diez and Hege, 1980). The $N_{min}$ method, developed by Wehrmann and Scharpf (1979), is basically a soil extraction with a NaCl + CaCl$_2$ solution that extracts soluble NO$_3^-$ and NH$_4^+$ and exchanges unspecifically bound NH$_4^+$. Generally, the NH$_4^+$ quantities recovered by the extraction are low in comparison with nitrate. Only in more acid soils and in soils having received a slurry dressing are higher amounts of NH$_4^+$ found.

Bronner (1974), in assessing the available N in sugar beet (*Beta vulgaris*) soils, extracts the soil with a NaCl solution, while the N mineralization potential is determined by an extraction of soil with boiling water (Bronner and Bachler, 1979). The data obtained by both extractions provided a satisfactory guide for the N fertilizer rate for sugar beets.

Fox and Piekielek (1978) found a highly significant correlation between the N extracted by a NaHCO$_3$ solution and the N uptake of corn, grown under field conditions. Roberts *et al.* (1980) reported that the NO$_3^-$ in the soil was well correlated with the N uptake and yield of *Zea mays*.

In flooded soils the most important N form directly available to plants is NH$_4^+$. According to Shiga and Ventura (1976), the exchangeable NH$_4^+$ in paddy soils is depleted to a very low level during 40 days after transplanting. Similar results were found by Keerthisinghe *et al.* (in press). From the total N taken up by the rice crop in many cases more than 50% originated from the pool of exchangeable + non-exchangeable NH$_4^+$, the rest from the organic N fraction. The contribution from the organic soil N to the total N uptake of the crop was higher during the dry than during the wet season (Keerthisinghe *et al.*, in press). Exchangeable NH$_4^+$ in flooded soils behaves like fertilizer N. Therefore, the level of exchangeable NH$_4^+$ at transplanting of rice can serve as a guide for the rating of N fertilizer, as has been shown recently by Mengel *et al.* (in press).

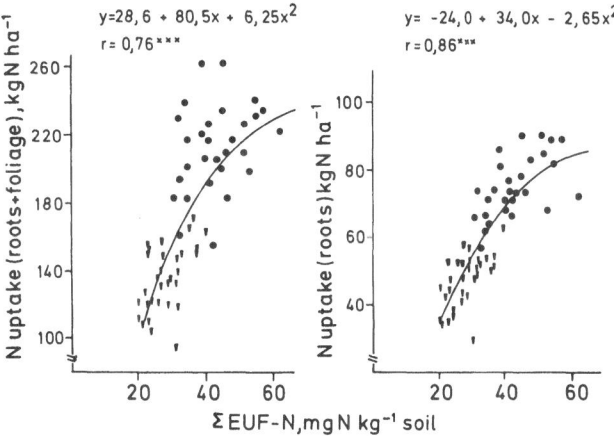

**Figure 10.** Relation between soil N extracted by EUF and N uptake of sugar beets, Σ EUF = inorganic + organic N. After Recke (1984). Used with permission.

Recent investigations have provided evidence that the electroultrafiltration method (EUF method) developed by Nemeth (1979) is also suitable for the determination of soil available N. Besides $NO_3^-$ and $NH_4^+$ a small quantity of organic N is also extracted by EUF (Nemeth et al., 1979). This quantity seems to be a N fraction that is easily mineralized by microbes. As shown in Figure 10, a highly significant correlation was obtained between the EUF-extracted soil N and the N uptake of sugar beets grown in the field on various luvisols derived from loess in the northern part of Germany (Recke, 1984). Particularly in sugar beet growing, the EUF soil test has gained in importance during recent years in Germany and Austria. In numerous field experiments Wiklicky et al. (1983) have shown a significant relationship between the EUF soil test for N and the increase in sugar yield by N fertilizer application.

# IV. Phosphorus

## A. Phosphate Fractions

In recent years valuable review articles on phosphate have been published: Dalal (1977b) on organic phosphates, Parfitt (1978) on phosphate adsorption, Khasawneh and Doll (1978) on rock phosphate, and Haynes (1982) on the effect of liming on phosphate availability. These papers should be consulted for more detailed information.

Principle phosphorus forms in soils are Ca-phosphates, adsorbed phosphates, occluded phosphates, and organic phosphates. The proportions of these fractions on total soil phosphate differ greatly for the various soil types. According to Walker and Syers (1976), during pedogenesis Ca-phosphates decrease and occluded phosphate increases. Thus, highly weathered acid soils are rich in occluded phosphate while their Ca-phosphate content is nil or almost nil. Organic phosphates may differ considerably according to the organic matter content of soils. In highly weathered podsolic soils, the proportion of organic phosphate may be as low as 5% of the total soil phosphate, while in humus-rich alpine soils 90% of total soil phosphate may be organic (Dalal, 1977b).

The proportion of Ca-phosphate increases with soil pH and Ca concentration in the soil (Schachtschabel and Heinemann, 1964). Thus, in calcareous soils the formation of Ca-phosphate tends to prevail. Solubility and availability differ considerably and follow the sequence: $CaHPO_4 \cdot 2H_2O$ (brushite) > $CaHPO_4$ > (monetite) > $Ca_8H_2(PO_4)_6 \cdot 5H_2O$ (octo-calciumphosphate) > $\beta\text{-}Ca_3(PO_4)_2$ ($\beta$-tricalciumphosphate) > $Ca_{10}(PO_4)_6H(OH)_2$ (hydroxyapatite) > $Ca_{10}(PO_4)_6F_2$(fluorapatite). Apatites may contain $CaCO_3$ in the lattice structure. Such impurities increase the apatite solubility considerably (Khasawneh and Doll, 1978). Transition of Ca-phosphates from one form to another one can be described by the following equations:

$$Ca(H_2PO_4)_2 + Ca^{2+} \rightleftarrows 2CaHPO_4 + 2H^+$$

$$6CaHPO_4 + 2Ca^{2+} \rightleftarrows Ca_8H_2(PO_4)_6 + 4H^+$$

$$Ca_8H_2(PO_4)_6 + 2Ca^{2+} + 2H_2O \rightleftarrows Ca_{10}(PO_4)_6(OH)_2 + 4H^+.$$

The reaction sequence shows that under high pH conditions and abundant $Ca^{2+}$ the formation of apatites is favored. A reverse reaction is true if apatites are exposed to low $Ca^{2+}$ and high $H^+$ concentration.

In mineral soils with pH below 7 the fraction of adsorbed phosphate is dominant (Larsen, 1967; Schachtschabel and Heinemann, 1964). Adsorption occurs at sesquioxides (Fe–Al-oxides), allophanes, clay minerals, organic Al–Fe-complexes, and calcite (Parfitt, 1978). In the case of non-specific adsorption, phosphate is adsorbed to protonated OH groups located at the surface of the above-mentioned particles. The bond thus formed depends on Coulomb forces and is therefore weak and greatly dependent on pH.

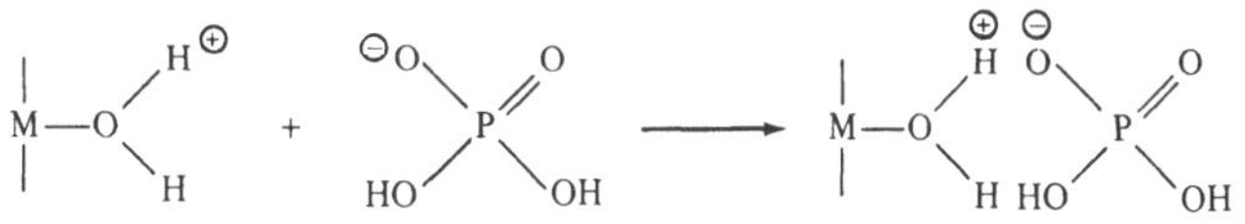

Since the process is related to protonated OH groups non-specific adsorption increases with a drop in soil pH.

Specific phosphate adsorption represents a ligand exchange between OH groups and phosphate according to the following equations:

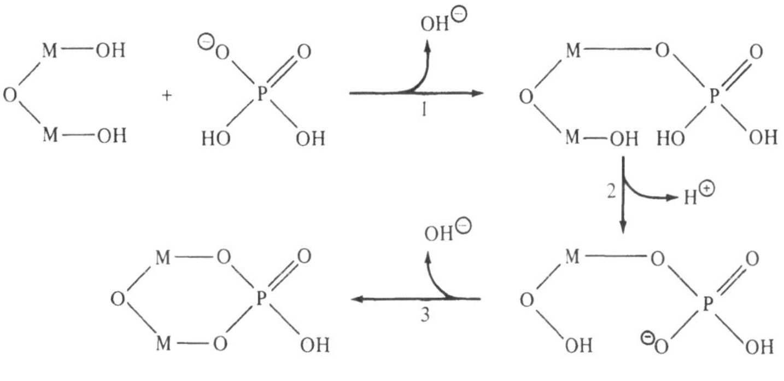

Reaction 1: $H_2PO_4^-$ exchanges with the $OH^-$ of the adsorbing surface. $OH^-$ is released. The bond established is mononuclear.

Reaction 2: The adsorbed phosphate is deprotonated even under low soil pH conditions.

Reaction 3: A second ligand exchange occurs that results in a further release of $OH^-$. Phosphate is bound with two bonds (binuclear adsorption), and a hexagonal structure is established (Hingston *et al.*, 1974).

Specific adsorption of one $H_2PO_4^{2-}$ in a binuclear type is, as shown above, associated with the net release of one $OH^-$. It is for this reason that the process is promoted by low pH. The binuclear bond is much stronger than the mononuclear one (Taylor and Ellis, 1978). Phosphate aging, which means the formation of more stable compounds, comprises the transition of mononuclear bound phosphate to binuclear one. By means of infrared spectra Nanzyo and Watanabe (1982) showed that goethite binds phosphate in a binuclear way in a pH range between 3.3 and 11.9, which demonstrates that even under very acid conditions stable binuclear phosphate adsorption occurs. The so-adsorbed phosphate still may dissociate $H^+$ from its last OH group. The p$K$ of such a binuclear goethite phosphate complex is 6.7 (Nanzyo and Watanabe, 1982).

In contrast to the non-specific adsorption, the specific adsorption is not only brought about by Coulomb forces, but chemical forces are also involved. The specific adsorption thus is also called chemi-adsorption. According to Bowden *et al.* (1977), the free energy for the adsorption process comprises Coulomb forces, chemical forces, and interactions.

$$\Delta G_{ad.} = \Delta G_{coul.} + \Delta G_{chem.} + \Delta G_{inter.}$$

Of particular interest is the adsorption of phosphate to calcite, a process that plays a major role in calcareous soils. It starts with the chemiadsorption and the formation of amorphous Ca-phosphates. These are gradually rendered in a crystalline structure (Parfitt, 1978). In systems with low phosphate concentration, apatites are finally formed in this way.

Adsorbed and dissolved phosphate tends to equilibrate. With an increase in phosphate concentration adsorption increases following a curve that represents approximately a Langmuir plot:

$$A = \frac{A_{max} K c}{1 + Kc} \qquad [8]$$

where $A$ is the amount of adsorbed phosphate, $A_{max}$ the maximum of adsorbed phosphate, $c$ the phosphate concentration of the equilibrated system, and $K$ the constant related to the strength of adsorption. In Figure 11 adsorption curves for five different soils are shown from the work of Okajima *et al.* (1983). Strongest adsorption was found in an ando soil, weakest in a brown lowland soil. Burnham and Lopez-Hernandez (1982), in investigating some hundred soil samples on their phosphate adsorption potential, found highest adsorption with inceptisols and virtually no phosphate adsorption in histosols. Phosphate adsoprtion of oxisols, generally believed to be strong phosphate fixers, was only high if the soils contained considerable sesquioxides. According to Schwertmann and

**Figure 11.** Phosphate concentration of the equilibrated solution in relation to adsorbed phosphate of five different soils. After Okajima *et al.* (1983), *Soil Science and Plant Nutrition* 29:271–283. Copyright © 1983 by the Japanese Society for Soil Science and Plant Nutrition. Used with permission.

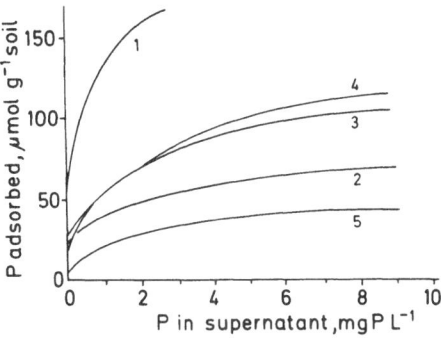

Schieck (1980), phosphate adsorption may also occur in calcareous soils, provided they are rich in Fe-oxides. The phosphate adsorption potential of sesquioxides is particularly high if they are in an amorphous status. With the transition from the amorphous to a more crystalline status, the phosphate adsorption potential decreases. Thus, according to Burnham and Lopez-Hernandez (1982) the phosphate adsorption potential of crystalline goethite and gibbsite is almost nil. Kawai (1980) found in ando soils a highly significant correlation between the phosphate adsorption potential and the content of amorphous Al-oxides. The specific surface of phosphate adsorbing particles has a great influence on adsorption as was shown by Lin *et al.* (1983b) for kaolinite, goethite, and gibbsite. The latter proved to be a particularly strong phosphate adsorber.

Phosphate adsorption can be affected by silicates (Scheffer *et al.*, 1980) and organic anions that compete with phosphates for adsorption sites. Strengite $[Fe^{III} H_2PO_4(OH)_2]$ and also variscite $[AlH_2PO_4(OH)_2]$ are only stable under acid conditions (Larsen, 1967) and therefore play no major role in arable soils. According to Ryden *et al.* (1973), neither phosphate type is formed in acid soils.

About 50% of the total organic phosphates in soils are phosphate esters of inositol. Besides hexaphosphate inositol, also di- tri- and tetra-inositol phosphates occur in soils (Dalal, 1977b). In comparison with the inositol phosphates, other organic soil phosphates such as nucleotides and phospholipids are present in low quantities and play no major role. Inositol phosphates may be adsorbed to soil colloids, and thus are impeded in their mobility. Organic phosphates can hardly be taken up by plants. Mobilization of organic phosphates is brought about by phosphatase, which splits off a phosphoryl group by hydrolysis, thus forming free inorganic phosphate. The enzyme phosphatase is present in various microorganisms (*Aspergillus, Penicillium, Mucor, Rhizopus, Bacillus*, and *Pseudomonas*), in the hyphae of the mycorrhiza, and can also be released by plant roots. Microorganisms feeding from organic C in the rhizosphere are particularly involved in the turnover of organic phosphates. Mineralization of organic phosphates increases with temperature and is generally higher in tropical soils than in soils of the moderate climate. According to Helal and Sauerbeck (1984), turnover of organic phosphates is particularly high in the rhizosphere.

### B. Phosphate Dynamics

Phosphate concentration of the soil solution is influenced by various processes: by the P uptake of plants, by the assimilation or release of inorganic phosphate from the pool of organic phosphate, and by equilibration with Ca-phosphates and with phosphate adsorption sites. These complicated dynamics are shown in Figure 12 in a simplified way. Soluble phosphate added to the soil finally will go to the strongest sink. Assuming a calcareous soil with a $Ca^{2+}$ concentration in the soil solution typical for such soils, conditions will favor the formation of apatite (Parfitt, 1978). Phosphate concentration equilibrated with apatite under neutral pH conditions is extremely low; for hydroxyapatites about $10^{-7}$ to $10^{-8}$mol $L^{-1}$. Thus, the process of apatite formation is a very strong sink for phosphate and may draw from the fraction of adsorbed phosphate if this is equilibrated with a higher phosphate concentration. Thus, Hooker et al. (1980a) found on five different soils with pH > 7 that P fertilizer application at first enriched the Fe–Al phosphate fraction (mainly adsorbed phosphate). In the following years this fraction decreased and the fraction of Ca-phosphates increased. Owing to the high pH of calcareous soils, phosphate adsorption is not very strong, which means that the phosphate concentration equilibrated with the adsorbed phosphate is relatively high and may be on the order of $10^{-4}$mol $L^{-1}$. Under such conditions there is a permanent flow of phosphate from the adsorbed

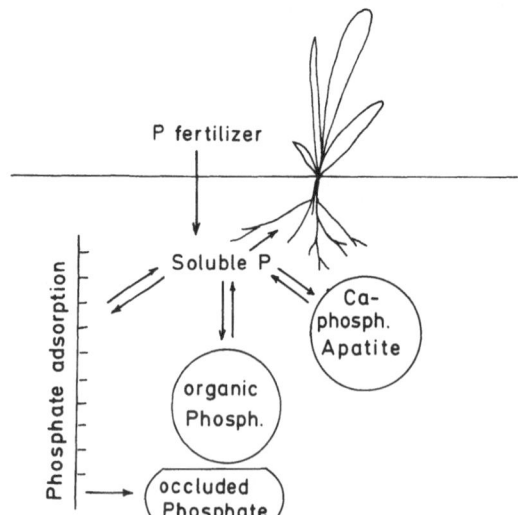

**Figure 12.** Soil phosphate fractions and their intrrelations.

fraction to apatite. Fertilizer phosphate added to such soils will at first be adsorbed or precipitated in a more soluble Ca-phosphate form, but finally the phosphate of these fractions will be shifted to apatite. Considering an acid soil rich in sesquioxides the process of adsorption is the strongest sink for phosphate, and hence fertilizer phosphate and phosphate equilibrated with more soluble Ca-phosphates as well as phosphate released from the organic pool will migrate to the adsorption sites. The process of adsorption is much quicker than the formation of apatites. Besides adsorption, phosphate occlusions also occur in such soils.

The basic difference between a system with apatite as the strongest sink and a system with adsorption sites as the strongest phosphate sink represents the fact that the phosphate concentration of the equilibrated solution is independent of the quantity of apatite present, whereas in the adsorption system the concentration of the equilibrated solution increases with an increase in the amount of adsorbed phosphate (see Figure 11). This is of relevance for phosphate uptake of plants. Too low a phosphate concentration ($< 10^{-5}$mol L$^{-1}$) may be the reason for an insufficient phosphate supply of plants even if the quantity of available phosphate is high.

Phosphate dynamics as shown schematically in Figure 12 may be influenced by various factors; by far the most important are H$^+$ and Ca$^{2+}$. Hence, Ca$^{2+}$ uptake of roots or H$^+$ secretion of roots may affect phosphate dynamics in the rhizosphere to a high degree. If in the system Ca-phosphates are dominating, Ca$^{2+}$ uptake and H$^+$ release of roots will promote phosphate solubilization. H$^+$ released in calcareous soils will be quickly neutralized. Since phosphates are involved in H$^+$ neutralization their solubility may also be increased according to the following reaction:

$$Ca_8H_2(PO_4)_6 + 4H^+ \rightarrow 6CaHPO_4 + 2Ca^{2+}$$

In soils with adsorbed phosphate as the principle fraction, pH decrease in the rhizosphere will depress the phosphate availability.

The question of how liming influences the phosphate availability deserves particular interest. According to the above-considered concepts, liming of acid soils should increase the phosphate availability. In soils rich in Al and Fe, an increase of the pH level may result in the formation of polynuclear complexes such as $Al_6(OH)_{15}^{3+}$ or $Al_{13}(OH)_{32}^{7+}$, which strongly adsorb phosphate. In such cases liming may decrease the phosphate availability (Haynes, 1982). A pH increase in the soil shifts the $HPO_4^{2-}/H_2PO_4$ ratio to a higher level, which also promotes adsorption. Sims and Ellis (1983), in studying the effect of liming on an acid Al-rich ultisol, found that phosphate availability was best increased with low lime rates that increased the soil pH only slightly from 3.37 to 3.61 (pH, KCl), but reduced the exchangeable Al considerably. Rhue and Hensel (1983) found on a sandy acid soil that liming decreased rather than increased the phosphate uptake of potatoes. The soil was low in Fe- and A1-oxides and therefore desorption of phosphate may have played a minor role. Hauter (1983) investigated the effect of pH increase from 5 to 7 (KC1) on the phosphate buffer curve with five tropical soils and one soil from Germany. All tropical soils showed typical phosphate adsorption curves. In four of the tropical soils investigated, the buffer curve shifted to a steeper slope with a rise in pH. This means that phosphate adsorption was favored by the higher pH. In one tropical soil the reverse was found; at pH 5 the buffer power curve was steeper than at pH 7. In the soil from Germany (luvisol derived from loess), an adsorption curve reflecting a Langmuir relationship was only found at pH 5, while at pH 7 the buffer curve, as shown in Figure 13, had a sigmoid shape. This shape is supposed to indicate that in addition to adsorption, phosphate precipitation was also involved. The curves were obtained from soil incubation with moisture content on the order of field capacity. The incubation time was 4 weeks. From the curves it can be derived that at the same level of phosphate quantity (y axis), e.g., 200 mg lactate-soluble P $kg^{-1}$ soil, the phosphate concentration at pH 7 (2.1 mmol $L^{-1}$) was more than twice as high as at pH 5 (0.8 mmol $L^{-1}$). Results of Hagemann and Müller (1976) are in good agreement with this observation. These authors found that a pH increase promoted the availability of adsorbed phosphate and a pH drop increased the availability of Ca-phosphate.

The equilibration of a soluble fertilizer incorporated into a soil is not only a question of adsorption and/or precipitation but also of diffusion. Even if a soluble phosphate is well mixed with the soil, the phosphate distribution is not homogeneous. There will be spots with high phosphate concentrations and low phosphate concentrations. As shown above, phosphate diffusion is a slow process, and hence an even distribution brought about by diffusion

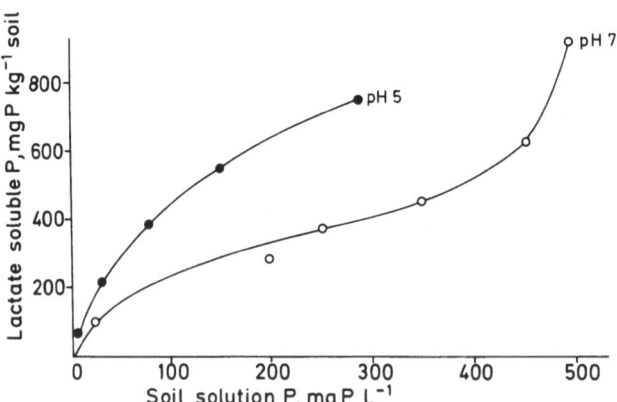

**Figure 13.** Relationship between the equilibrated-phosphate concentration of the solution and the lactate-soluble phosphate of a luvisol at pH 5 and at pH 7. Modified after Hauter (1983). Used with permission.

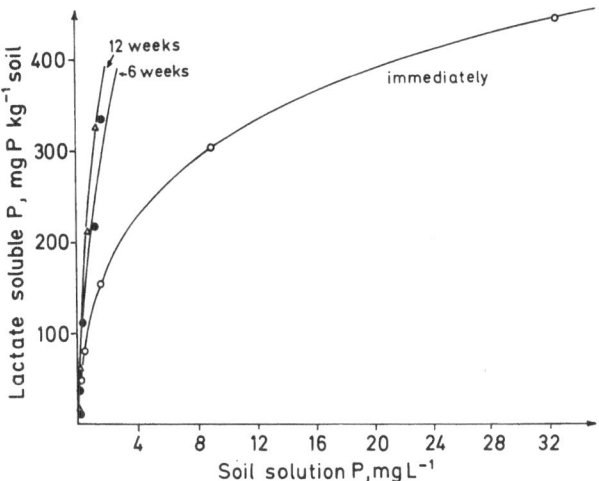

**Figure 14.** Phosphate buffer curves of an acid soil obtained immediately after phosphate application and 6 and 12 weeks after phosphate application. After Barekzai (1984). Used with permission.

lasts some time. This contact time between phosphate fertilizer and soil has an impact on fertilizer availability (Barrow and Shaw, 1975) and thus is of practical importance. Figure 14 shows the phosphate buffer curves of an acid soil (pH 4.5) established directly after phosphate application and after an incubation time of 6 and 12 weeks (Barekzai, 1984). The slope of the curve obtained directly after application was much flatter than the slopes obtained 6 and 12 weeks later. During this time soil water was kept at field capacity. The data on the x axis (intensity) relate to the phosphate concentration of the soil solution; the data on the y axis (quantity) to the phosphate extracted by Ca-lactate. This extractant recovers more or less the isotopically exchangeable phosphate (Keerthisinghe and Mengel, 1979). As is evidenced by Figure 14, the solubility of the fertilizer phosphate decreased considerably during 6 weeks. It is supposed that this decrease originated from a more even distribution of phosphate, which means that phosphate migrated to the strongest adsorption sites. In addition, the transition from a mononuclear to a binuclear adsorption will also have contributed to the decrease in solubility. During the incubation period, a significant pH increase was observed, which shows that a ligand exchange, phosphate for $OH^-$, was involved (Barekzai, 1984). Besides the remarkable decrease in phosphate concentration, the content of lactate-soluble phosphate was also considerably diminished. Since under the low pH conditions a formation of less soluble Ca phosphates is unlikely, phosphate occlusion may have occurred during the period of incubation.

Barekzai (1984), in studying ten different soils on their effect of aging phosphate fertilizer, found that in all soils the Ca lactate-soluble and the electroultrafiltration-extractable phosphate decreased significantly during a period of 6 months. This decrease in availability was clearly reflected in the P uptake of *Lolium perenne* grown in a pot experiment. Interestingly enough, the lowest aging effect was found in a soil with a $CaCO_3$ content of 67% and a flat phosphate adsorption curve, while the highest aging effect was found in an acid soil with a steep adsorption curve (soil shown in Figure 14). It is suggested that phosphate aging is associated with phosphate adsorption and possibly also with phosphate occlusion.

Phosphate aging is of practical relevance since it has a strong impact on fertilizer phosphate recovery. According to the above-mentioned results, aging of fertilizer phosphate was especially high in the acid soil. This is in agreement with results of Sturm and Isermann (1978) who evaluated various arable soils and grassland in Germany on their phosphate fertilizer recovery. They found a recovery of about 80% in neutral soils with a good structure and also in grassland. Poorest recovery of 50% was found on dry soils with low pH.

Occluded phosphate can be recovered under reducing soil conditions. Reduction of $Fe^{III}$-oxides to $Fe^{2+}$ breaks the Fe–Al-oxide skin of occluded phosphates so that phosphate is released (Ponnamperuma, 1972). The reaction of phosphate fertilizer with soils depends much on solubility of the fertilizer. Water-soluble phosphates will relatively quickly equilibrate with the soil, while with less soluble phosphate fertilizer the process will proceed slowly. A particular problem occurs if fertilizers contain apatite such as phosphate rock and partially acidulated phosphates. Their solubility in soils decreases with an increase in soil pH. They are therefore well suited for acid soils, while in soils with a pH $> 5.5$ (KC1) their effect is sporadic and often nil (Khasawneh and Doll, 1978; van der Paauw, 1965; Amberger and Gutser, 1976; Hammond *et al.*, 1980; Obigbesan and Mengel, 1981b; Judel *et al.*, 1982; Palmer and Gilkes, 1983).

As considered in section II, phosphate concentration of the soil solution and phosphate buffer power are most important factors of phosphate availability. Both factors can be well assessed by establishing a buffer curve in which the phosphate concentration of the soil solution (intensity) appears on the x axis and the quantity of available phosphate appears on the y axis. The latter can be obtained by the determination of isotopically exchangeable phosphate, approximately also by extraction with an acid soil test extractant. In order to get a realistic value of the phosphate concentration, it is pertinent to establish the curve in a moist soil (about field capacity) and not in a soil suspension. Phosphate adsorption studied in soil suspension may provide reliable information about the nature of adsorption (Ozanne and Shaw, 1967), the transformation of the obtained curves on soil systems, however, meets with difficulties. As has been shown by Barrow and Shaw (1975) and by Lin *et al.* (1983a), phosphate concentration in soil suspensions depends on the solution soil ratio. Even the kind of shaking

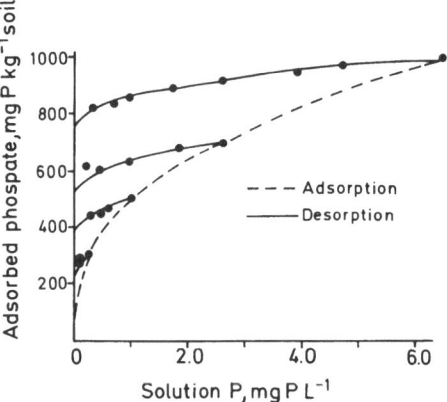

**Figure 15.** Hysteresis of phosphate adsorption. Relationship between the phosphate concentration of the equilibrated solution and adsorbed phosphate at the adsorption and desorption process. Modified after Barrow and Shaw (1975), *Soil Science* 119:311–320. Copyright © 1975 by The Williams & Wilkins Co., Baltimore. Used with permission.

may affect phosphate adsorption, since a vigorous shaking may result in a break of soil particles and thus in the exposure of new phosphate adsorption sites. It thus appears that phosphate incubation in a moist soil provides a more realistic picture with regard to phosphate adsorption and phosphate concentration in the soil solution because an incubation time of some weeks is required to obtain equilibrium (see Figure 14). Curves thus obtained reflect the phosphate concentration of the bulk soil solution. The steepness of the curve at a given phosphate concentration reflects the buffer power. There exists tremendous differences in the buffer power of soils. Generally acid tropical soils have a high buffer power, which means that at a given quantity of available phosphate, the phosphate concentration of the soil solution is relatively low (Pagel and van Huay, 1976; Obigbesan and Mengel, 1981a). Barekzai (1984) in investigating soils widely differing in phosphate buffer power found that at the same quantity of available phosphate, the phosphate concentration of the soil solution can differ by a factor of 100.

Buffer curves of acid soils show on the y axis mainly adsorbed phosphate. The curve may differ depending on whether it is obtained by adsorption or desorption. Desorption may be characterized by a considerable hysteresis as shown in Figure 15 from the work of Barrow and Shaw (1975). A similar finding was reported by Okajima *et al.* (1983). Hysteresis is of relevance for phosphate availability, since at the same quantity of phosphate (y axis), phosphate concentration of the equilibrated solution will be much lower with the desorption process than with the adsorption process.

The quantity of phosphate dissolved in the soil solution is low; even in fertile soils it is generally on the order of 1 kg P ha$^{-1}$ in the upper soil layer (0 to 0.3 m). Nevertheless, this dissolved phosphate is of high importance for the rate of phosphate uptake by the crop, as shown in section II. Under vigorous growth conditions the dissolved phosphate is quickly depleted and must be continously replenished by the available phosphate pool. As was considered above, this replenishment depends on the phosphate buffer power. The quantity of phosphate that equilibrates with the dissolved phosphate generally is more than 100 times higher than the quantity of the soil solution phosphate. Virtually all soil test methods provide no information about the most important factors of phosphate availability, phosphate concentration in the soil solution and the phosphate buffer power. Also with extractants such as bicarbonate (Olsen et al., 1954) or water (van der Paauw, 1971), the phosphate quantities recovered are more than 10 to 100 times higher than the phosphate present in the soil solution. Thus all methods, especially those with acid extractants, extract quite a considerable amount of phosphate from the fraction of the so-called labile pool (Larsen, 1967). In some cases, acid extractants even dissolve phosphate that under field conditions will not be available to plants. Thus, it has been shown that the application of apatite containing phosphate fertilizers may increase the lactate-soluble phosphate considerably, because this acid extractant also dissolves apatites, which generally under field conditions were not available (Werner, 1969; Gutser and Amberger, 1976). It is for this reason that in Central Europe the CAL method was introduced. In this method, the extractant consists of Ca-lactate, Ca-acetate, and acetic acid, pH 4.1 (Schüller, 1969). Often the relationship between the phosphate status of the soil assessed by a routine soil test and phosphate fertilizer response is unsatisfactory (Schachtschabel, 1980). It is feasible that the relationship could be improved if besides the soil test the phosphate buffer power would be taken into consideration as has been recently shown by Nair and Mengel (1984). Sibbesen (1983) in evaluating a number of soil test methods came to the conclusion that the methods which extract only low quantities of phosphate (water, bicarbonate) were superior to acid and buffer extractants. To the former group also belongs the electroultrafiltration method (EUF), which is closely related to the water extract method (Nemeth, 1979). Best correlations between P uptake and phosphate soil tests were obtained by the anion exchange resin method (Sibbesen, 1983) as described by Amer et al. (1955) and Sibbesen (1978). This technique simulates a root, and thus it takes into consideration the phosphate intensity (concentration) and the phosphate buffer power.

# V. Potassium

## A. Potassium Fractions in the Soil

Review papers on potassium were published by Rich (1968, 1972) on K in minerals, by Schroeder (1974) on K fractions in soils, and by Mengel and Kirkby (1980) on potassium in crop production. These papers may be consulted for more detailed information in the particular field of interest.

Potassium status of soils depends greatly on soil parent material and its degree of weathering (Graham and Fox, 1971). Highly weathered soils and also histosols are generally low in K, while young soils derived from materials rich in K-bearing minerals contain abundant K. The most important K-bearing minerals are alkali feldspars (3 to 12% K), Ca–Na feldspars (0 to 3% K), and micas (biotite, 5 to 8% K; muscovite, 6 to 9% K). From the secondary clay minerals vermiculite and illite may also contain K on the order of 1 to 6%. Weathering of these minerals is associated with the release of $K^+$. The minerals formed during the weathering of micas correspond with the following sequence (Schroeder, 1976): micas (~10% K) → hydromicas (6 to 8% K) → illite (4 to 6% K) → transition minerals (~3% K) → vermiculite or smectite (~2% K). Weathering of feldspars results in the formation of a Si–Al–O skin on the surface of the mineral, impeding the process of mineral degradation and therefore also of $K^+$ release (Rich, 1972). Consequently, with an advance in weathering the rate of $K^+$ release decreases.

Considering the mobility of soil K, different K fractions can be distinguished: $K^+$ in the lattice of minerals, interlayer $K^+$ between the tetrahedral layers of minerals such as micas, $K^+$ adsorbed to inner and outer surfaces of soil colloids, and $K^+$ dissolved in the soil solution. The latter is the smallest fraction and amounts to about 10 to 20 kg K ha$^{-1}$ in the top layer (0 to 0.3 m) of fertile soils. The adsorbed K is on the order of 300 to 1000 kg K, the interlayer K of $10^3$ to $10^4$ kg K, the total K of $2 \times 10^4$ to $5 \times 10^4$ kg K ha$^{-1}$ in the top layer of mineral soils. The smallest fractions, the $K^+$ in the soil solution and the adsorbed $K^+$, are of paramount importance for the $K^+$ supply of plants. Both tend to equilibrate with each other. The fraction of adsorbed $K^+$ is not clear cut, especially in cases where 2:1 clay minerals with interlayer $K^+$ are involved. In Figure 16 different $K^+$

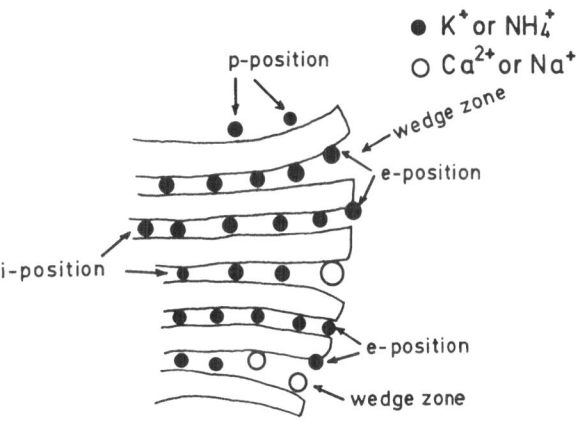

**Figure 16.** Scheme of $K^+$ adsorption positions and wedge zones of weathered mica.

adsorption positions are shown for a mica with marginal expansion zones (wedge zones). Potassium ions adsorbed to the planar surface (p-position) are unspecifically adsorbed, $K^+$ still has its hydration shell, the bond is relatively weak, and the equilibrated $K^+$ concentration is high. Potassium ions of the interlayer (i-position) are very strongly bound and are in a dehydrated status. They equilibrate with a very low $K^+$ concentration. Potassium bound at the edges of the interlayer (e-position) take an intermediary position, which means that the strength of the bond and the equilibrated $K^+$ concentration are not very high (Shouwenburg and Schuffelen, 1963; Ehlers *et al.*, 1968).

These three different $K^+$ positions can be characterized by the Gapon coefficient (Bolt *et al.*, 1963),

$$\frac{K^+ \text{ ads}}{Ca^{2+}\text{ads}} = k \ \frac{K_s^+}{\sqrt{Ca_s^{2+}}} \qquad [9]$$

where $K^+$ ads $Ca^{2+}$ ads is the concentration of adsorbed $K^+$ and $Ca^{2+}$, $K_s^+$, $Ca_s^{2+}$ the concentration of $K^+$ and $Ca^{2+}$ in the equilibrated contact solution, and $k$ the Gapon coefficient.

From the equation it can be derived that the K bond strength will increase with increasing values of $k$. Correspondingly, p-positions are characterized by a low value ($< 10$), e-positions by a medium value (about 100), and i-positions by a very high value ($> 1000$). Considering a highly $K^+$-saturated illite, the equilibrated $K^+$ concentration of the contact solution is controlled by the $K^+$ in p-position and is thus comparatively high. The removal of $K^+$ from the solution results in a desorption of $K^+$. At first the $K^+$ of the p-positions is desorbed, but as soon as this fraction is depleted the e-positions will release $K^+$ and finally the $K^+$ of the i-positions (interlayer $K^+$) will also be desorbed. The difference in the strength of bonds for $K^+$ is evidenced by the $K^+$ adsorption curve. Figure 17 shows such curves from the work of Grimme *et al.* (1971). The clay of the sandy soil had a relatively high $K^+$ saturation, which means that a high proportion of $K^+$ was bound to p-positions and hence the curve is flat. The curve of the clay soil can be divided into two linear components, one with a steeper and one with a flatter slope. The latter runs approximately parallel to the curve of the sandy soil and represents the p-positions, while the steeper part is mainly related to the e-positions. Both curves also represent the $K^+$ buffer power of the soils in question.

$K^+$ adsorption sites of kaolinites, organic matter, planar outer surfaces of clay minerals, and to some extent also adsorption sites of inner surfaces of smectite are less specific and correspond in their strength to p-positions. In soils in which these weak $K^+$ adsorption bonds dominate, the $K^+$ is very mobile and can easily be replaced by other cation species, such as $Ca^{2+}$, $Mg^{2+}$, $H^+$ and is thus prone to leaching.

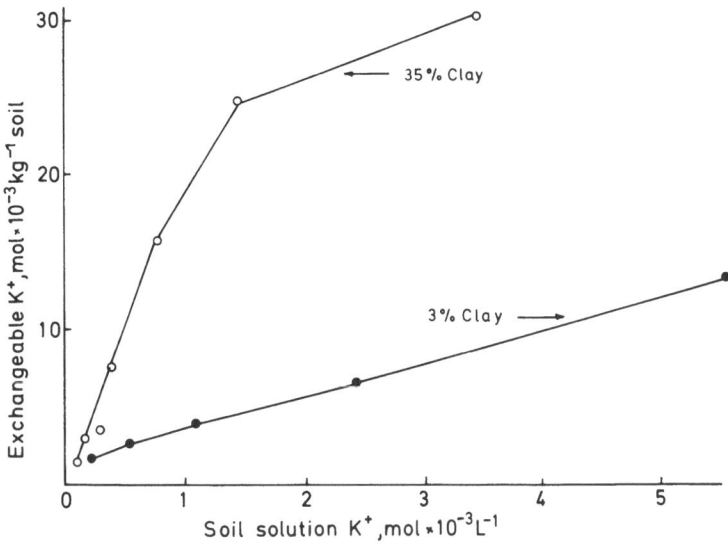

**Figure 17.** Relationship between the equilibrated $K^+$ concentration of the solution and the exchangeable $K^+$ of two soils with different clay content. After Grimme *et al.* (1971), *Landw. Forsch. SH* 26(I):165–176. Copyright © 1971 by J.D. Sauderländer's Verlag. Used with permission.

The so-called exchangeable $K^+$ relates to the replacement with $NH_4^+$. This $K^+$ fraction consists mainly of the $K^+$ from the p- and e-positions. The surplus of $NH_4^+$ present in the extractant brings about a contraction of the wedge zones so that the release of interlayer $K^+$ is blocked. Potassium extraction by means of electroultrafiltration mainly recovers the K of the p-positions (Grimme, 1982). The interlayer $K^+$ belongs to the so-called non-exchangeable fraction, although it can be exchanged by a repeated treatment with $Na^+$ or $Ca^{2+}$ (Jackson and During, 1979).

## B. Potassium Dynamics and Availability

Potassium in the soil solution and also the $K^+$ of the p-positions can be easily absorbed by plant roots and can also be translocated vertically into deeper soil layers by rainfall. The $K^+$ of the e-positions is less mobile and $K^+$ of the interlayer is released only under particular conditions. This fraction may contribute to plant nutrition and thus deserves particular interest. The tetrahedral layers of micas, illites, and vermiculites bear a relatively high negative charge, and the fact that the unhydrated $K^+$ fits well into the hexagonal network of the tetrahedral layer is the reason why $K^+$ is specifically and strongly adsorbed between the layers of 2:1 minerals. The same is true for $NH_4^+$ but not for $Na^+$. The hydration energy for $Na^+$ is 400 k J mol$^{-1}$ and 315 k J mol$^{-1}$ for K$^+$. Since dehydration is a prerequisite of specific adsorption between the tetrahedral layers, K$^+$ is much more prone to adsorption than Na$^+$. This is the reason for the preferential adsorption of K$^+$ to micas, illites, and vermiculites in comparison with Na$^+$, Mg$^{2+}$, and Ca$^{2+}$.

Release of K$^+$ from interlayers is an exchange and diffusion process (von Reichenbach, 1972). Diffusion depends largely on the expansion of the mineral and therefore on soil moisture; exchange depends on the cation species and their concentrations near the surface of the mineral. Net release of K$^+$ will only occur if the K$^+$ concentration of the adjacent solution is low. Martin and Sparks (1983), in studying the release of non-exchangeable K$^+$ from a sandy loam and a loamy sand by shaking the soil with a H$^+$ charged ion exchanger, found a substantial net release of K$^+$ with a K$^+$ concentration in the contact solution of about 1 to 2 µmol L$^{-1}$. This is a concentration level to which a plant root may deplete the soil solution level in the rhizosphere. According to Martin and Sparks (1983), the K$^+$ net release can be described by a first-order diffusion process:

$$d\, K_t/dt = k_2 (K_0 - K_t),$$                      [10]

where $K_0$ is total non-exchangeable K, $K_t$ is released K at time $t$, and $k_2$ is release rate coefficient of non-exchangeable K.

From this equation it follows that the rate of K$^+$ release decreases with the amount of K$^+$ already desorbed and increases with the time of contact. Net release depends also on the K$^+$ concentration of the contact solution, the K$^+$ release rate being higher at lower contacting K$^+$ concentrations (Mortland, 1958). The critical K$^+$ concentration of the outer solution at which no net release occurs differs for the various minerals. Biotite, owing to its trioctahedral structure, showed a net release at a relatively high K$^+$ concentration of about 1 mmol L$^{-1}$, while for muscovite the net release of K$^+$ was already depressed at a concentration of 0.1 mmol K$^+$ L$^{-1}$ (Rausell-Colom et al., 1965). Siimilar results were reported by Scott and Smith (1966). They found that at a K$^+$ concentration of 0.2 mmol L$^{-1}$, biotite still was capable of releasing K$^+$, while muscovite and illite were not. If there was a permanent disequilibrium between the contact solution and the mineral, almost all interlayer K$^+$ could be exchanged from the minerals studied with the exception of illite, of which only 50% of the total interlayer K$^+$ was desorbed. The velocity of K$^+$ release, however, differed considerably between the various minerals. For the release of 90% of the total interlayer K$^+$ the following durations were required: vermiculite 30 minutes, phlogopite (Mg biotite) 1.5 days, biotite 3 days, and muscovite 315 days. This may demonstrate that muscovite (K mica), although rich in K$^+$, is much inferior to biotite in supplying plants with K$^+$. The higher resistance of muscovite to weathering is related to its dioctahedral structure. Jackson and During (1979) in investigating the K$^+$ release of six different soils with vermiculite, illite, and smectite as the main clay minerals found critical $K^+$ concentrations in the contact solution in the order of 0.1 to 2 mmol $L^{-1}$.

Experiments of Newman (1969) also showed that the critical concentration at which the $K^+$ release stops is much lower for muscovite than for biotite. The $K^+$ concentrations were independent of the quantity of $K^+$ still present in the interlayers. Thus, interlayer $K^+$ behaved like a precipitate equilibrated with its solubility product. $K^+$ release was associated with a pH increase of the contact solution that did not originate from the adsorption of $H^+$ but was related to oxidation of $Fe^{II}$ according to the equation

$$4\ Fe^{II} + 4\ H^+ + O_2 \rightarrow 4\ Fe^{III} + 2\ H_2O.$$

From this it follows that $H^+$ promotes the oxidation of lattice Fe and this in turn may be the reason why low pH conditions favor $K^+$ release (Newman, 1969).

Release of interlayer $K^+$ depends also on the exchanging cation species. Calcium and $Na^+$, when replacing $K^+$ in the interlayer, bring about an expansion of the mineral that in turn will favor the $K^+$ release. Ammonium and $H^+$, when exchanging for interlayer $K^+$, do not expand the mineral and may thus impede $K^+$ diffusion and release (Scott and Smith, 1966). Ammonium adsorbed to edge positions of the mineral may bring about a contraction and thus hamper the release of interlayer $K^+$. The favorable effect of $Ca^{2+}$ and $Na^+$ on $K^+$ release was reported by Jackson and During (1979) who pretreated the soils with $Ca^{2+}$ and $Na^+$ before measuring the release of $K^+$. High release rates were found especially when the treatment was carried out at a higher temperature (50°C).

There is considerable evidence in the literature that plants may feed from interlayer $K^+$. Mortland et al. (1956) were able to demonstrate that biotite may be altered to vermiculite by continuous $K^+$ removal by plants. Malquori et al. (1975) found that wheat was able to exploit the $K^+$ of biotite. Schroeder and Dümmler (1966) reported that in soils of northern Germany interlayer $K^+$ of illites and illitic transition minerals play a major role in supplying crops with non-exchangeable $K^+$. On the alluvial soils of the Punjab about 80 to 90% of total K taken up by corn or wheat resulted from interlayer $K^+$ (Singh and Brar, 1977). On these young mica-rich soils, although intensively cropped, K fertilizer application did not result in yield increase of corn or wheat. The capability of exploiting interlayer $K^+$ differs between plant species. Steffens and Mengel (1979) found that ryegrass (*Lolium perenne*) could feed from interlayer $K^+$ for a longer period without yield depression, while red clover (*Trifolium pratense*) could not. These authors hold the view that *Lolium*, because of its greater root length as compared with *Trifolium*, may still grow satisfactorily at a relatively low $K^+$ concentration at which *Trifolium* will already suffer from $K^+$ deficiency (Steffens and Mengel, 1981). It is feasible that the difference in root mass, root length, and root morphology between monocots and dicots is the reason why monocots feed better from interlayer $K^+$ than dicots. This assumption is in good agreement with the general finding that monocots do not respond to K fertilizer application as much as dicots (van der Paauw, 1958; Schön et al., 1976).

Interlayer $K^+$ is not a permanent everlasting $K^+$ source for plants. The source may be exhausted as was shown by Mengel and Wiechens (1979) with permanent cropping of Lolium perenne in pot experiments. The first seven clippings showed no significant yield differences between treatments with or without K fertilizer, although in the treatment without $K^+$ plants mainly fed from interlayer $K^+$. At the eighth clipping the yield of the treatment without $K^+$ was significantly lower than the yield obtained with $K^+$, and at the eleventh clipping the yield of the $K_0$ treatment was only about 40% of the yield with $K^+$ application. Obviously with an increase in depletion of interlayer $K^+$, the $K^+$ release rate decreased and eventually did not meet the crop requirement. In the example cited above (Mengel and Wiechens, 1978), only 2.5% of the total soil $K^+$ had been released when heavy $K^+$ deficiency symptoms were observed in the plants.

Higher losses of interlayer $K^+$ result in an increase of the $K^+$ fixation potential. Losses from interlayer $K^+$ may occur when soils are cropped for a longer time without $K^+$ fertilizer application (Nielsen, 1970). Minerals may also have lost $K^+$ during pedogenesis, especially when the soil was transported by water (Niederbudde, 1967). Loss of interlayer $K^+$ results in the formation of wedge zones, which when contacted with higher $K^+$ concentrations bind $K^+$ followed by a contraction of the expanded zone. In this way, fertilizer $K^+$ may be fixed. The proportion of wedge zones decreases for various minerals in the following sequence: micas > illite > vermiculite > montmorillonite (Arifin and Tan, 1973). Normal fertilizer rates in the range of 100 to 300 kg K ha$^{-1}$ can be completely fixed (Burkart and Amberger, 1978). Higher rates, however, resulted in remarkable yield increases (Schäfer and Siebold, 1972). Fixation could be overcome by heavy K rates on the order of 1000 to 2000 kg K ha$^{-1}$. As soon as the interlayer $K^+$ was refilled, normal K fertilizer rates could be applied.

Under field conditions, equilibration of fertilizer $K^+$ with the $K^+$ depleted interlayer positions is not a quick process but may last some months (Amberger et al., 1974). This was also evidenced by Karbachsch (1978), who treated a heavy clay alluvium soil with a high K fixation potential with increasing $K^+$ fertilizer rates. Potassium concentration measured in the soil solution increased steeply with the increase in fertilizer $K^+$ after a relatively short contact time of 30 days. After 90 days of contact, the curve had become flat, indicating that the bulk of solution K had been adsorbed. This finding is of practical importance, since it shows that on K-fixing soils fertilizer K is readily available shortly after application but declines in its availability with time. Hence on $K^+$-fixing soils K fertilizer should be applied at sowing or even top dressed. Assessment of available $K^+$ is mostly carried out by treating the soil with an extractant. The quantity of $K^+$ thus obtained depends on the nature of the extractant. Extraction with $H_2O$, e.g., the saturation extract, recovers mainly the dissolved $K^+$ and thus reflects the $K^+$ concentration of the soil solution. Extraction with $NH_4^+$ as the replacing cation relates mainly to the so-called exchangeable $K^+$, while acid extractants may desorb a considerable proportion of interlayer $K^+$. With these techniques a K quantity is obtained that gives no reliable information about the $K^+$ intensity ($K^+$ concentration) in a soil. Both the saturation extract method and the EUF technique provide data that are more closely related to the $K^+$ concentration of the soil solution which, as outlined above, is an important factor of K availability. It may be for this reason that frequently the $K^+$ concentration of the soil solution corresponds better with the $K^+$ uptake of crops than the exchangeable $K^+$ (von Braunschweig and Mengel, 1972; During and Duganzich, 1979). In addition, the EUF-extractable $K^+$ was a better indicator of $K^+$ availability than the exchangeable $K^+$ in field studies with flooded rice (Wanasuria et al., 1981). Besides the $K^+$ concentration the $K^+$ buffer power should also be considered when estimating $K^+$ availability. Thus, During and Duganzich (1979) obtained highly significant correlations ($r^2 = 0.9$) for the K uptake of white clover grown on eight soils widely differing in cation exchange capacity when $K^+$ concentration and $K^+$ buffer power of soils were taken into consideration.

A further approach in assessing the $K^+$ availability of soils represents the determination of the activity ratio for $K^+$ and $Ca^{2+}$ as proposed by Beckett (1964). The activity ratio $a_k/\sqrt{a_{Ca}}$ at which neither adsorption nor desorption of $K^+$ occurs is the $AR_0$ value. It is supposed to be an important indicator of $K^+$ availability. Results obtained until now, however, do not support this concept (Wild et al., 1969; During and Duganzich, 1979). Nair and Grimme (1979), growing oats (Avena sativa) on four different soils, found only a weak correlation between the $AR_0$ data and the $K^+$ uptake of oats while the correlation with the EUF data was satisfactory ($r^2 = 0.77$). The $AR_0$ value is a ratio and thus not only dependent on the $K^+$, but also on the $Ca^{2+}$ concentration. Calcium, however, has no major impact on K availability (Mengel, 1963). This may be one reason why $AR_0$ is not a satisfactory indicator of $K^+$ availability. A further drawback of the technique is that the activity ratio is obtained from a soil suspension. It therefore does not reflect the actual $K^+$ and $Ca^{2+}$ concentrations in the soil solution. The $AR$ value is also affected by the soil solution ratio (During and Duganzich, 1979).

Woodruff (1955) in an attempt to determine the change of free anergy for the K–Ca exchange postulated the following formula:

$$\Delta F = RT \ln \frac{a_k}{a_{Ca}^{1/2}} \qquad [11]$$

where $a_k$ is the activity of $K^+$ and $a_{ca}$ the activity of $Ca^{2+}$ in soil solution. The formula is derived from the equilibrium constant for the exchange system "$K \rightarrow Ca$" and therefore does not relate to the free-energy change as such but to the standard free energy ($\Delta G^0$). Woodruff (1955) assumed the activities of the adsorbed $Ca^{2+}$ and $K^+$ as one, and for this reason these were not taken into consideration. It is questioned, however, whether such an assumption is justified since the "activities" of adsorbed $Ca^{2+}$ and $K^+$ can differ considerably with regard to the activities of $Ca^{2+}$ and $K^+$ in solution, depending on the type of adsorbing surfaces. It is thus doubted whether the above formula, which simply represents the K/Ca activity ratio, will be of major use for estimating soil $K^+$ availability. In contrast to Woodruff's (1955) approach, in a recent investigation Jardine and Sparks (1984) considered all four reaction partners in studying Ca–K exchange. Exchanging $Ca^{2+}$ for $K^+$ at planar sites yielded positive values for $\Delta G^0$, while negative values were obtained when $Ca^{2+}$ bound at interlayer positions was replaced by $K^+$. Obviously this approach is suitable for gaining a deeper insight into the energy change of cation exchange processes in soils.

## VI. Concluding Remarks

A major factor in efficient fertilizer use is to minimize fertilizer losses. Nitrogen losses may occur in the form of gaseous nitrogen ($NH_3$, $N_2$, $N_2O$) and by leaching, particularly $NO_3^-$. The losses depend largely on soil conditions, climatic conditions, and even on the crop. Therefore, no specific recommendation can be given on how to minimize such losses. According to the nitrogen dynamics as discussed in section III, type of N fertilizer, time and rate of application, and, in particular cases, the use of denitrification inhibitors will contribute to the improvement of N fertilizer efficiency.

The efficient use of phosphate fertilizers can be affected by phosphate occlusion and phosphate adsorption, especially by binuclear binding. Particularly acid soils are prone to these types of phosphate aging. In many cases an increase in soil pH may reduce the rate of phosphate fixation or occlusion. Phosphate application at sowing and band application as well as the use of the appropriate phosphate type may help to improve the recovery of fertilizer phosphate.

Potassium losses are mainly due to leaching and thus occur on coarse-textured soils under humid conditions. Split application of K fertilizer adjusted to the crop demand and to weather conditions are considered a useful measure to improve the efficiency of K fertilizers under conditions where leaching losses may occur.

As cropping intensity increases and the use of fertilizers becomes greater, it becomes more important to control soil fertility by soil tests. Until now the various techniques used have not often yielded satisfactory results. As has been shown in the previous sections, nutrient availability depends greatly on soil types and crops, especially on their rooting systems. The reliability of soil tests will likely be improved if further soil and plant factors are taken into consideration, such as the nutrient buffer power of soils, the rooting depth, and the rooting characteristics of the crops in question. Routine techniques for estimating nitrogen fertility status of soils are badly needed. New techniques and approaches, as pointed out in section III, should encourage researchers and technicians to develop routine methods for application on a broad scale of soil types and crops. The implementation of this objective would make a significant contribution to the economic and ecological use of fertilizer.

# References

Aguilars, A., and A. van Diest. 1981. Rock phosphate mobilization induced by the alkaline uptake pattern of legumes utilizing symbiotically fixed nitrogen. *Plant and Soil* 61:27–42.

Aichberger, K. 1982. Veränderungen des pflanzenverfügbaren Bodenstickstoffgehaltes ($N_{min}$) im Jahresablauf. *Die Bodenkultur* 33:277–288.

Allison, F.E., M. Kefauver, and E.M. Roller. 1953. Ammonium fixation in soils. *Soil Sci. Soc. Am. Proc.* 17:107–110.

Amberger, A., and R. Gutser. 1976. Aussagekraft von Bodenuntersuchungsmethoden in langjährigen Feldversuchen mit verschiedenen P-Formen. *Landw. Forsch. SH* 33(I):18–38.

Amberger, A., and K. Vilsmeier. 1979. Hemmung der Nitrifikation des Güllestickstoffs durch Dicyandiamid. *Z. Acker- Pflanzenbau* 148:239–246.

Amberger, A., R. Gutser, and K. Teicher. 1974. Kaliumernährung der Pflanzen und Kaliumdynamik auf Kalium-fixierendem Boden. *Plant and Soil* 40:269–284.

Amberger, A., K. Vilsmeier, and R. Guster. 1982. Stickstofffraktionen verschiedener Güllen und deren Wirkung im Pflanzenversuch. *Z. Pflanzenern. Bodenk.* 145:325–336.

Amer, F., D.R. Bouldin, C.A. Black, and F.R. Duke. 1955. Characterization of soil phosphorus by anion exchange resin adsorption and $^{32}$P-equilibration. *Plant and Soil* 6:391–408.

Andrew, C.S., and A.D. Johnson. 1976. Effect of calcium, pH and nitrogen on the growth and chemical composition of some tropical and temperate pasture legumes. II. Chemical composition (calcium, nitrogen, potassium, magnesium, sodium and phosphorus). *Austr. J. Agr. Res.* 27:625–636.

Antoniw, L.D., and J.I. Sprent. 1978. Primary metabolites of *Phaseolus vulgaris* nodules. *Phytochemistry* 17:675–678.

Araragi, M., and B. Tangcham. 1979. Effect of rice straw on the composition of volatile soil gas and microflora in the tropical paddy field. *Soil Sci. Plant Nutr.* 25(3):283–295.

116                                                                          Konrad Mengel

Arifin, H.F., and K.H. Tan. 1973. Potassium fixation and reconstitution of micaceous structures in soils. *Soil Sci.* 116:31–35.

Arnon, I. 1969. Transition from extensive to intensive agriculture in Israel with fertilizers. In: *Transition from extensive to intensive agriculture with fertilizers.* pp. 13–24. Int. Potash Inst., Berne.

Bachthaler, G., and A. Wagner. 1973. Ergebnisse langjähriger Vergleichsversuche Stroh-Gründüngung und Strohverbrennung unter verschiedenen Standortbedingungen. *Bayer. Ldw. Jahrb.* 50:436–461.

Barber, S.A. 1962. A diffusion mass flow concept of soil nutrient availability. *Soil Sci.* 93:39–49.

Barber, S.A. 1974. Influence of the plant root on ion movement in soils. In: E.W. Carson (ed.), *The plant root and its environment.* pp. 525–564. University Press of Virginia, Charlottesville.

Barber, S.A., and J.H. Cushman. 1981. Nitrogen uptake model of agronomic crops. In: J.K. Iscander (ed.), *Modelling waste water rennovation land treatment.* pp. 382–409. Wiley Interscience, New York.

Barber, S.A., J.M. Walker, and E.H. Vasey. 1963. Mechanisms for the movement of plant nutrients from the soil and fertilizer to the plant root. *Agr. Food Chem.* 11:204–207.

Barekzai, A. 1984. Alterung von wasserlöslichem Phosphat—untersucht in Gefäß- und Modellversuchen. Ph. D. Thesis, FB 19, Justus Liebig-University Giessen.

Barrow, N. J., and T.C. Shaw. 1975. The slow reactions between soil and anions. 5. Effects of period of prior contact on the desorption of phosphate from soils. *Soil Sci.* 119:311–320.

Bauer, W.D. 1981. Infection of legumes by Rhizobia. *Ann. Rev. Plant Physiol.* 32:407–449.

Beck, T. 1983. Die N-Mineralisierung von Böden im Laborbrutversuch. *Z. Pflanzenern. Bodenk.* 146:243–252.

Beckett, P.H.T. 1964. Studies on soil potassium. II. The "immediate" Q/I relations of labile potassium in the soil. *J. Soil Sci.* 15:9–23.

Berthelin, J., and C. Leyval. 1982. Ability of symbiotic and non-symbiotic rhizospheric microflora of maize (*Zea mays*) to weather micas and to promote plant growth and plant nutrition. *Plant and Soil* 68:369–377.

Bhat, K.K.S., and P.H. Nye. 1974. Diffusion of phosphate to plant roots in soil. III. Depletion around onion roots without root hairs. *Plant and Soil* 41:383–394.

Biederbeck, V.O., C.A. Campbell, K.E. Bowren, and R.N. McIver. 1980. Effect of burning cereal straw on soil properties and grain yields in Saskatchewan. *Soil Sci. Soc. Am. J.* 44:103–111.

Blair, G.J., and O.W. Boland. 1978. The release of phosphorus from plant material added to soil. *Austr. J. Soil Res.* 16:101–111.

Bole, J.B. 1973. Influence of root hairs in supplying soil phosphorus to wheat. *Can. J. Soil Sci.* 53:169–175.

Bolt, G.H., M.E. Sumner, and A. Kamphorst. 1963. A study of the equilibria between the categories of potassium in an illite soil. *Soil Sci. Soc. Am. Proc.* 27:294–299.

Borst, N.P., and C. Mulder. 1971. Stikstofgehalte, Stikstofbemesting en opbrengst van wintertarwe op zeezand-, kleien zavelgronden in Nord-Holland. *Betrijfsontwikkeling* 2:31–36.

Bowden, J.W., A.M. Posner, and J.P. Quirk. 1977. Ionic adsorption on variable change mineral surfaces. Theoretical-charge development and titration curves. *Austr. J. Soil Res.* 15:121–136.

Breisch, H., A. Guckert, and O. Reisinger. 1975. Etude au microscope électronique de la zone apicale des racines des mais. *Soc. Bot. Fr. Coll. Rhizosphere* 122:55–60.

Bremner, J.M. 1959. Determination of fixed ammonium in soil. *J. Agr. Sci.* 52:147–160.

Bremner, J.M., and D.W. Nelson. 1968. Chemical decomposition of nitrite in soils. *9th Int. Congr. Soil Sci. Trans.* (Adelaide) 2:495–503.

Breteler, H., and A.L. Smith. 1974. Effect of ammonium nutrition on uptake and metabolism of nitrate in wheat. *Neth. J. Agr. Sci.* 22:73–81.

Broadbent, F.E., and T. Nakashima. 1971. Effect of added salt on nitrogen mineralization in three California soils. *Soil Sci. Soc. Am. Proc.* 35:457–460.

Bronner, H. 1974. Der leichtlösliche Stickstoff im Boden im Zusammenhang mit Kenndaten der Rübenentwicklung. *Landw. Forsch. SH* 30(II):39–44.

Bronner, H., and W. Bachler. 1979. Der hydrolysierbare Stickstoff als Hilfsmittel für die Schätzung des Stickstoffnachlieferungsvermögens von Zuckerrübenböden. *Landw. Forsch.* 32:255–261.

Buresh, R.J., M.E. Casselman, and W.H. Patrick, Jr. 1980. Nitrogen fixation in flooded soil systems, a review. *Adv. Agron.* 33:149–192.

Burkart, N., and A. Amberger. 1978. Einfluß der Kaliumdüngung auf die Verfügbarkeit des Kaliums in K-fixierenden Böden im Verlaufe der Vegetationszeit. *Z. Pflanzenern. Bodenk.* 141:167–179.

Burnham, C.P., and D. Lopez-Hernandez. 1982. Phosphate retention in different soil taxonomic classes. *Soil Sci.* 134:376–380.

Campbell, C.A. 1978. Soil organic carbon, nitrogen and fertility. In: M. Schnitzer and S.U. Khan (eds.), *Developments in soil science 8: soil organic matter.* pp. 174–271. Elsevier, Amsterdam.

Chin, W.-T., and W. Kroontje. 1963. Urea hydrolysis and subsequent loss of ammonia. *Soil Sci. Soc. Am. Proc.* 27:316–318.

Claassen, N., and S.A. Barber. 1976. Simulation model for nutrient uptake from soil by growing plant root system. *Agron. J.* 68:961–964.

Claassen, N., and A. Jungk. 1982. Kaliumdynamik im wurzelnahen Boden in Beziehung zur Kaliumaufnahme von Maispflanzen. *Z. Pflanzenern. Bodenk.* 145:513–525.

Claassen, N., K. Hendriks, and A. Jungk. 1981. Rubidium-Verarmung des wurzelnahen Bodens durch Maispflanzen. *Z. Pflanzenern. Bodenk.* 144:533–545.

Coic, Y., C. Lessaint, and F. Le Roux. 1962. Effects de lat nature ammoniacale ou nitrique de l'alimentation azotée et du changement de la nature de cette alimentation sur le métabolisme des anions et cations chez la tomate. *Ann. Physiol. Veg.* 4:117–125.

Dalal, R.C. 1977a. Fixed ammonium and carbon-nitrogen ratios of some Trinidad soils. *Soil Sci.* 124:323–327.

Dalal, R.C. 1977b. Soil organic phosphorus. *Adv. Agron.* 29:83–117.

Daly, G.T. 1966. Nitrogen fixation by nodulated *Alnus rugosa*. *Can. J. Bot.* 44:1607–1621.

Dejaegere, R., and L. Neirinckx. 1978. Proton extrusion and ion uptake: some characteristics of the phenomenon in barley seedlings. *Z. Pflanzenphysiol.* 89:129–140.

Delwiche, C.C. 1983. Cycling of elements in the biosphere. In: A. Läuchli and R.L. Bieleski (eds.), *Inorganic plant nutrition. Encycl. plant physiol.* New Series, Vol. 15. pp. 212–238. Springer-Verlag, New York.

Diez, T., and U. Hege. 1980. Stickstoffdüngung des Weizens nach Bodenuntersuchung ($N_{min}$) in Abhängigkeit von den Standortverhältnissen. *Bayer. Ldw. Jahrb.* 57:944–951.

Dijkshoorn, W. 1957. A note on the cation-anion relationships in perennial ryegrass. *Neth. J. Agr. Sci.* 5:81–85.

Döbereiner, J., J.M. Day, and P.J. Dart. 1972. Nitrogenous activity and oxygen sensitivity of the *Paspalum notatum-Azotobacter paspali* association. *J. Gen. Microbiol.* 71:103–116.

Drew, M.C., P.H. Nye, and L.V. Vaidyanathan. 1969. The supply of nutrient ions by diffusion to plant roots in soil. I. Absorption of potassium by cylindrical roots of onion and leek. *Plant and Soil* 30:252–270.

During, C., and D.M. Duganzich. 1979. Simple empirical intensity and buffering capacity measurements to predict potassium uptake by white clover. *Plant and Soil* 51:167–176.

Ehlers, W., H. Gebhardt, and B. Meyer. 1968. Untersuchungen über die positions-spezifische Bindung des Kaliums an Illit, Kaolinit, Montmorillonit und Humus. *Z. Pflanzenern. Bodenk.* 119:173–186.

Ellenberg, H. 1964. Stickstoff als Standortfaktor. *Ber. Dtsch. Bot. Ges.* 77:82–92.

Farquhar, G.D., P.M. Firth, R. Wetselaar, and B. Weir. 1980. On the gaseous exchange of ammonia between leaves and the environment: determination of the ammonia compensation point. *Plant Physiol.* 66:710–714.

Feigenbaum, S., and K. Mengel. 1979. The effect of reduced light intensity and suboptimal potassium supply on $N_2$ fixation and N turnover in *Rhizobium* infected lucerne. *Physiol. Plant.* 45:245–249.

Fox, R.H., and W.P. Piekielek. 1978. A rapid method for estimating the nitrogen-supplying capability of a soil. *Soil Sci. Soc. Am. J.* 42:751–753.

Führ, F., and J.M. Bremner. 1964. Untersuchungen zur Fixierung des Nitritstickstoffs durch die organische Masse des Bodens. *Landw. Forsch. SH* 18:43–51.

Ganry, F., G. Guiraud, and Y. Dommergues. 1978. Effect of straw incorporation on the yield and nitrogen balance in the sandy soil–pearl millet cropping system of Senegal. *Plant and Soil* 50:647–662.

Gaskell, J.F., A.M. Blackmer, and J.M. Bremner. 1981. Comparison of effects of nitrate, nitrite, and nitric oxide on reduction of nitrous oxide to dinitrogen by soil microorganisms. *Soil Sci. Soc. Am. J.* 45:1124–1127.

Graff, O., and H. Kühn. 1977. Einfluß des Regenwurms *Lumbricus terrestris* L. auf die Ertrags- und Nährstoffwirkung einer Strohdüngung. *Landw. Forsch.* 30:86–93.

Graham, E.R., and R.L. Fox. 1971. Tropical soil potassium as related to labile pool and calcium exchange equilibria. *Soil Sci.* 111:318–322.

Graham, J.H., R.T. Leonard, and J.A. Menge. 1981. Membrane mediated decrease in root exudation responsible for phosphorus inhibition of vesicular–arbuscular mycorrhiza formation. *Plant Physiol.* 68:548–552.

Grimme, H. 1982. K desorption in an external electrical field as related to clay content. *Plant and Soil* 64:49–54.

Grimme, H., K. Nemeth, and L.C. von Braunschweig. 1971. Beziehungen zwischen dem Verhalten des Kaliums im Boden und der K-Ernährung der Pflanze. *Landw. Forsch. SH* 26(I):165–176.

Guckert, A., H. Breisch, and O. Reisenauer. 1975. Etude au microscope électronique des relations mucigel-argile-microorganismes. *Soil Biol. Biochem.* 7:241–250.

Guo, P.-C., J. Bohring, and H.W. Scherer. 1983. Verhalten von Dünger-$NH_4^+$ in Böden unterschiedlicher tonmineralischer Zusammensetzung. *Z. Pflanzenern. Bodenk.* 146:752–759.

Gutser, R., and A. Amberger. 1976. Aussagekraft einiger Bodenuntersuchungsmethoden für Phosphat in ein- und mehrjährigen Gefäßversuchen. *Landw. Forsch. SH* 33(I):39–51, Kongreßband.

Gutser, R., and K. Teicher. 1976. Veränderungen des lösliche Stickstoffes einer Ackerbraunerde unter Winterweizen im Jahresverlauf. *Bayer. Ldw. Jahrb.* 53:215–226.

Hagemann, O., and S. Müller. 1976 Untersuchungen über den Einfluß des pH-Wertes auf die Ausnutzung von Düngerphosphaten und die Mobilisierung von Bodenphosphaten. *Arch. Acker- Pflanzenbau Bodenk.* 20:805–815.

Haider, K., and Farooq-e-Azam. 1983. Umsetzung $^{14}$C markierter Pflanzeninhalts-stoffe im Boden in Gegenwart von $^{15}$N-Ammonium. *Z. Pflanzenern. Bodenk.* 146:151–159.

Hale, M.G., and L.D. Moore. 1979. Factors affecting root exudation II 1970–1978. *Adv. Agron.* 31:93–124.

Hammond, L.L., S.H. Chien, and J.R. Polo. 1980. Phosphorus availability from partial acidulation of two rock phosphates. *Fert. Res.* 1:37–49.

Hauck, R.D. 1971. Quantitative estimates of nitrogen-cycle processes: concepts and review. In: *Nitrogen-15 in soil–plant studies.* pp. 65–80. IAEA, Vienna.

Hauter, R. 1983. Phosphatmobilisierung in Abhängigkeit vom pH des Bodens unter besonderer Berücksichtigung der Rhizosphäre. Ph. D. Thesis, FB 19, Justus Liebig-University Giessen.

Haynes, R.J. 1982. Effects of liming on phosphate availability in acid soils. *Plant and Soil* 68:289–308.

Helal, H.M., and D.R. Sauerbeck. 1984. Influence of plant roots on carbon and phosphorus metabolism in soil. *Plant and Soil* 76:175–182.

Hendriks, L., N. Claassen, and A. Jungk. 1981. Phosphatverarmung des wurzel-nahen Bodens und Phosphataufnahme von Mais und Raps. *Z. Pflanzenern. Bodenk.* 144:486–499.

Hiltner, L. 1904. Über neuere Erfahrungen und Probleme auf dem Gebiet der Bodenbakteriologie unter besonderer Berücksichtigung der Gründüngung und Brache. *Arb. Deutsch. Landw. Ges.* 98:59–79.

Hingston, F.J., A.M. Posner, and J.P. Quirk. 1974. Anion adsorption by goethite and gibbsite. II. Desorption of anions by hydrous oxide surfaces. *J. Soil Sci.* 25:16–26.

Hinman, W.C. 1964. Fixed ammonium in some Saskatchewan soils. *Can. J. Sci.* 44:151–157.

Holford, I.C.R. 1976. Effects of phosphate buffer capacity of soil on the phosphate requirements of plants. *Plant and Soil* 45:433–444.

Hooker, M.L., G.A. Peterson, D.H. Sander, and L.A. Daigger. 1980a. Phosphate fractions in calcareous soils as altered by time and amounts of added phosphate. *Soil Sci. Soc. Am. J.* 44:269–277.

Hooker, M.L., D.H. Sander, G.A. Peterson, and L.A. Daigger. 1980b. Gaseous N losses from winter wheat. *Agron. J.* 72:789–792.

Israel, D.W., and W.A. Jackson. 1978. The influence of nitrogen nutrition on ion uptake and translocation by leguminous plants. In: C.S. Andrew and E.J. Kamprath (eds.), *Mineral nutrition of legumes in tropical and subtropical soils.* pp. 113–128. CSIRO, Australia.

Itoh, S., and S.A. Barber. 1983. Phosphorus uptake by six plant species as related to root hairs. *Agron. J.* 75:457–461.

Jackson, B.L.J., and C. During. 1979. Studies of slowly available potassium in soils of New Zealand. I. Effects of leaching, temperature and potassium depletion on the equilibrium concentration of potassium in solution. *Plant and Soil* 51:197–204.

Jardine, P.M., and D.L. Sparks. 1984. Potassium–calcium exchange in a multireactive system. II. Thermodynamics. *Soil Sci. Soc. Am. J.* 48:45–50.

Judel, G.K., W.G. Gebauer, and K. Mengel. 1982. Einfluß der Löslichkeit verschiedener Phosphatdüngemittel auf die Phosphataufnahme und den Ertrag von Sommerweizen. *Z. Pflanzenern. Bodenk.* 145:296–303.

Karbachsch, M. 1978. Kaliumernährung des Tabaks auf einem K-fixierenden nordwestiranischen Boden. *Z. Pflanzenern. Bodenk.* 141:513–522.

Kawai, K. 1980. The relationship of phosphorus adsorption to amorphous aluminum for characterizing andosols. *Soil Sci.* 129:186–190.

Keeney, D.R., and J.M. Bremner. 1966. Comparison and evaluation of laboratory methods of obtaining an index of soil nitrogen availability. *Agron. J.* 58:498–503.

Keerthisinghe, G., and K. Mengel. 1979. Phosphatpufferung verschiedener Böden und ihre Veränderung infolge Phosphatalterung. *Mitt. Dtsch. Bodenk. Ges.* 29:217–230.

Keerthisinghe, G., S.K. DeDatta, and K. Mengel. In press. Importance of exchangeable and nonexchangeable soil $NH_4^+$ in nitrogen nutrition of wetland rice. *Soil Sci.*

Keerthisinghe, G., K. Mengel, and S.K. DeDatta. 1984. The release of nonexchangeable ammonium ([15]N labelled) in wetland rice soils. *Soil Sci. Soc. Am. J.* 48:291–294.

Khasawneh, F.E., and E.C. Doll. 1978. The use of phosphate rock for direct applications to soils. *Adv. Agron.* 30:159–206.

Kirkby, E.A., and K. Mengel. 1967. Ionic balance in different tissues of the tomato plant in relation to nitrate, urea or ammonium nutrition. *Plant Physiol.* 42:6–14.

Kjellerup, V., and A. Dam Kofoed. 1983. Nitrogen fertilization in relation to leaching of plant nutrients from soil. Lysimeter experiments with [15]N. *Tidsskr. Planteavl.* 87:1–22.

Kowalenko, C.G., and D.R. Cameron. 1978. Nitrogen transformations in soil-plant systems in three years of field experiments using tracer and non-tracer methods of an ammonium-fixing soil. *Can. J. Soil Sci.* 58:195–208.

Larsen, S. 1967. Soil phosphorus. *Adv. Agron.* 19:151–206.

Legg, J.O., and F.E. Allison. 1959. Recovery of [15]N tagged nitrogen from ammonium-fixing soil. *Soil Sci. Soc. Am. Proc.* 23:131–134.

Lemon, E., and R. van Houtte. 1980. Ammonia exchange at the land surface. *Agron. J.* 72:876–883.

Lewis, D.G., and J.P. Quirk. 1967. Phosphate diffusion in soil and uptake by plants. III. [31]P-movement and uptake by plants as indicated by [32]P-autoradiography. *Plant and Soil* 26:445–453.

Liao, C.F.H., and W.V. Bartholomew. 1977. Nitrate absorption and transport by corn plants in soil system under different moisture regimes. In: *Proc. Int. sem. soil env. and fertility management in intensive agricult.* pp. 625–633. Soc. Sci. Soil and Manure, Japan. Nippon Dojohiryo Gakkai, Tokyo.

Lin, C., W.J. Busscher, and L.A. Douglas. 1983a. Multifactor kinetics of phosphate reactions with minerals in acidic soils. I. Modeling and simulation. *Soil Sci. Soc. Am. J.* 47:1097–1103.

Lin, C., H.L. Motto, L.A. Douglas, and W.J. Busscher. 1983b. Multifactor kinetics of phosphate reactions with minerals in acid soils. II. Experimental curve fitting. *Soil Sci. Soc. Am. J.* 47:1103–1109.

Liu, Zhi-yu, and Sheng-wu Qin. 1981. The study of nitrogen distribution around rice rhizosphere. In: *Proc. symp. paddy soil.* Institute of Soil Science, Academica Sinica. pp. 511–546. Springer-Verlag, Berlin, Heidelberg, New York.

Low, A.J., and E.R. Armitage. 1970. The composition of the leachate through cropped and uncropped soils in lysimeters compared with that of the rain. *Plant and Soil* 33:393–411.

Malquori, A., G. Ristori, and V. Vidrrich. 1975. Biological weathering of potassium silicates. I. Biotite. *Agrochimica* 19:522–529.

Martin, H.W., and D.L. Sparks. 1983. Kinetics of nonexchangeable potassium release from two coastal plain soils. *Soil Sci. Soc. Am. J.* 47:883–887.

Matar, A.E., J.L. Paul, and H. Jenny. 1967. Two-phase experiments with plants growing in phosphate-treated soil. *Soil Sci. Soc. Am. Proc.* 31:235–237.

Mederski, H.J., J. Stackhouse, and J.H. Wilson. 1960. Relation of soil moisture to ion absorption by corn plants. *Soil Sci. Soc. Am. Proc.* 24:149–152.

Mengel, K. 1963. Untersuchungen über das 'Kalium-Kalzium-potential.' *Z. Pflanzenern. Düngung Bodenk.* 103:99–111.

Mengel, K. 1983. Responses of various crop species and cultivars to fertilizer application. *Plant and Soil* 72:305–319.

Mengel, K., and L.C. von Braunschweig. 1972. The effect of soil moisture upon the availability of potassium and its influence on the growth of young maize plants (*Zea mays* L.). *Soil Sci.* 114:142–148.

Mengel, K., and R. Busch. 1982. The importance of the potassium buffer power on the critical potassium level in soils. *Soil Sci.* 133:27–32.

Mengel, K., and H. Casper. 1980. Der Einfluß der Bodenfeuchte auf die Verfügbarkeit von Nitratstickstoff im Boden. *Z. Pflanzenern. Bodenk.* 143:617–626.

Mengel, K., and E.A. Kirkby. 1980. Potassium in crop production. *Adv. Agron.* 33:59–110.

Mengel, K., and N. Malissiovas. 1982. Light dependent proton excretion by roots of entire vine plants (*Vitis vinivera* L.). *Z. Pflanzenern. Bodenk.* 145:261–267.

Mengel, K., and H.W. Scherer. 1981. Release of nonexchangeable (fixed) ammonium under field conditions during the growing season. *Soil Sci.* 131:226–232.

Mengel, K., and D. Steffens. 1982. Beziehung zwischen Kationen/Anionen-Aufnahme von Rotklee und Protonenabscheidung der Wurzeln. *Z. Pflanzenern. Bodenk.* 145:229–236.

Mengel, K., and M. Viro. 1978. The significance of plant energy status for the uptake and incorporation of $NH_4$ nitrogen by young rice plants. *Soil Sci. Plant Nutr.* 24(3):407–416.

Mengel, K., and B. Wiechens. 1979. Die Bedeutung der nicht austauschbaren Kaliumfraktion des Bodens für die Ertragsbildung von Weidelgras. *Z. Pflanzenern. Bodenk.* 142:836–847.

Mengel, K., P. Robin, and L. Salsac. 1983. Nitrate reductase activity in shoots and roots of maize seedlings as affected by the form of nitrogen nutrition and the pH of the nutrient solution. *Plant Physiol.* 71:618–622.

Mengel, K., H.G. Schön, G. Keerthisinghe, and S.K. DeDatta. In press. Importance of exchangeable and nonexchangeable ammonium for rice growth and grain yields of flooded soils. *Fert. Res.*

Mikkelsen, D.S., S.K. DeDatta, and W.N. Obcemea. 1978. Ammonia volatilization losses from flooded rice soils. *Soil Sci. Soc. Am. J.* 42:725–730.

Moghimi, A., M.E. Tate, and J.M. Oades. 1978. Phosphate dissolution by rhizosphere products. II. Characterization of rhizosphere products especially α ketogluconic acid. *Soil Biol. Biochem.* 10:283–286.

Mojallali, H., and S.B. Weed. 1978. Weathering of micas by mycorrhizal soybean plants. *Soil Sci. Soc. Am. J.* 42:367–372.

Moore, A.W. 1969. *Azolla*: biology and agronomic significance. *Bot. Rev.* 35:17–34.

Mortland, M.M. 1958. Kinetics of potassium release from biotite. *Soil Sci. Soc. Am. Proc.* 22:503–508.

Mortland, H.M., K. Lawton, and G. Uehara. 1956. Alteration of biotite to vermiculite by plant growth. *Soil Sci.* 82:477–481.

Müller, S., H. Ansorge, O. Hageman, H. Görlitz, J. Garz, and H. Stumpe. 1976. Untersuchungen über die Möglichkeiten einer Bemessung der ersten N-Gabe zu Getreide durch Berücksichtigung des Gehaltes an anorganischem Stickstoff im Boden. *Arch. Acker- Pflanzenbau Bodenk.* 20:713–722.

Munns, D.N. 1968. *Medicago sativa* in solution culture. III. Effects of nitrate on root hairs and infection. *Plant and Soil* 29:33–47.

Myers, R.J.K., and E.A. Paul. 1971. Plant uptake and immobilization of [15]N-labelled ammonium nitrate in a field experiment with wheat. In: *Nitrogen-15 in soil–plant studies*. pp. 55–64. IAEA, Vienna.

Nair, K.P.P., and K. Mengel. 1984. The importance of the phosphate buffer power for the phosphate uptake of rye. *Soil Sci. Soc. Am. J.* 48:92–95.

Nair, P.K., and H. Grimme. 1979. Q/I relations and electroultrafiltration of soils as measures of potassium availability to plants. *Z. Pflanzenern. Bodenk.* 142:87–94.

Nanzyo, M., and Y. Watanabe. 1982. Diffuse reflectance infrared spectra and ion adsorption properties of the phosphate surface complex on goethite. *Soil Sci. Plant Nutr.* 28:359–368.

Nemeth, K. 1979. The availability of nutients in the soil as determined by electroultrafiltration (EUF). *Adv. Agron.* 31:155–188.

Nemeth, K., I.Q. Makhdum, K. Koch, and H. Beringer. 1979. Determination of categories of soil nitrogen by electroultrafiltration (EUF). *Plant and Soil* 53:445–453.

Newman, A.C.D. 1969. Cation exchange properties of micas. I. The relation between mica composition and potassium exchange in solutions of different pH. *J. Soil Sci.* 20:357–373.

Neyra, C.A., and J. Döbereiner. 1977. Nitrogen fixation in grasses. *Adv. Agron.* 29:1–38.

Niederbudde, E.A. 1967. Mineralverwitterung und Kaliumfixierung in Anfangsstadien der Bodenbildung des oberen Etschtales. *Z. Pflanzenern. Düngung Bodenk.* 115:28–43.

Nielsen, J.D. 1970. Fixation and release of potassium in Danish soils. *Tidsskr. Planteavl.* 74:24–43.

Nishizawa, N., T. Yoshida, and Y. Arima. 1983. Electron microscopic study of associative $N_2$-fixing bacteria in roots of rice seedlings. *Soil Sci. Plant Nutr.* 29:261–270.

Nyatsanga, T., and W.H. Pierre. 1973. Effect of nitrogen fixation by legumes on soil acidity. *Agron. J.* 65:936–940.

Nye, P.H. 1966. The effect of the nutrient intensity and buffering power of a soil, and the absorbing power, size and root hairs of a root, on nutrient absorption by diffusion. *Plant and Soil* 25:81–105.

Nye, P.H. 1979. Diffusion of ions and uncharged solutes in soils and clays. *Adv. Agron.* 31:225–272.

Obigbesan, G.O., and K. Mengel. 1981a. Relationship between electroultrafiltration (EUF) extractable phosphate, P-uptake and P buffer capacity of selected tropical soils. *Niger. J. Soil Sci.* 1:1–12.

Obigbesan, G.O., and K. Mengel. 1981b. Use of electroultrafiltration (EUF) method for investigating the behaviour of phosphate fertilizers in tropical soils. *Fert. Res.* 2:169–176.

Okajima, H., H. Kubota, and T. Sakuma. 1983. Hysteresis in the phosphorus sorption and desorption processes of soils. *Soil Sci. Plant Nutr.* 29:271–283.

Olsen, S.R., and F.S. Watanabe. 1970. Diffusive supply of phosphorus in relation to soil texture variations. *Soil Sci.* 110:318–327.

Olsen, S.R., C.V. Cole, F.S. Watanabe, and L.A. Dean. 1954. Estimation of available phosphorus in soils by extraction with sodium bicarbonate. *USDA Cir. No. 939.*

Osborne, G.J. 1976. The significance of intercalary ammonium in representative surface and subsoils from Southern New South Wales. *Austr. J. Soil Res.* 14:381–388.

Overrein, K.N., and P.G. Moe. 1967. Factors affecting urea hydrolysis and ammonia volatilization in soil. *Soil Sci. Soc. Am. Proc.* 31:57–61.

Ozanne, P.G., and T.C. Shaw. 1967. Phosphate sorption by soils as a measure of the phosphate requirement for pasture growth. *Austr. J. Agr. Res.* 18:601–612.

Page, M.B., and O. Talibudeen. 1977. Nitrate concentrations under winter wheat and in fallow soil during summer at Rothamsted. *Plant and Soil* 47:527–540.

Pagel, H., and H. van Huay. 1976. Wichtige Parameter der Phosphat-Sorptionskurven einiger Böden der Tropen und Subtropen und ihre zeitliche Veränderung durch P-Düngung. *Arch. Acker- Pflanzenbau Bodenk.* 20:765–778.

Palmer, B., and R.J. Gilkes. 1983. The influence of application rate on the relative effectiveness of calcined Christmas Island C-grade rock phosphate and superphosphate when applied as mixtures. *Fert. Res.* 4:45–50.

Parfitt, R.L. 1978. Anion adsorption by soils and soil materials. *Adv. Agron.* 30:1–50.

Parthier, B. 1978. Die biologische Fixierung des atmosphärischen Stickstoffs. *Biol. Rdsch.* 16:345–364.

Paul, R.E., and R.L. Jones. 1976. Studies on the secretion of maize root cap slime. IV. Evidence for the involvement of dictyosomes. *Plant Physiol.* 57:249–256.

Peterson, W.R., and S.A. Barber. 1981. Soybean root morphology and K uptake. *Agron. J.* 73:316–319.

Pfaff, C. 1963. Das Verhalten des Stickstoffs im Boden nach langjährigen Lysimeterversuchen. *Z. Acker- Pflanzenbau* 117:77–99.

Ponnamperuma, F.N. 1972. The chemistry of submerged soils. *Adv. Agron.* 24:29–96.

Rausell-Colom, J.A., T.R. Sweatman, C.B. Wells, and K. Norish. 1965. Studies in artificial weathering of micas. In: *Experimental pedology proc.* pp. 40–70. Univ. Nottingham, 11th Easter School Agr. Sci.

Recke, H. 1984. Kalium- und Stickstoffverfügbarkeit südniedersächsischer Stand-orte-bestimmt mittels Elektro-Ultrafiltration (EUF) in Beziehung zu Ertrag und Qualität der Zuckerrübe. Ph. D. Thesis, FB 19, Justus Liebig-University Giessen.

Reddy, K.R., and P.S.C. Rao. 1983. Nitrogen and phosphorus fluxes from a flooded organic soil. *Soil Sci.* 136:300–307.

Reddy, K.R., W.H. Patrick, and R.E. Phillips. 1976. Ammonium diffusion as a factor in nitrogen loss from flooded soils. *Soil Sci. Soc. Am. J.* 40:528–533.

Renger, M., and O. Strebel. 1976. Nitratanlieferung an die Pflanzenwurzel als Funktion der Tiefe und der Zeit. *Landw. Forsch. SH* 33(II):13–19.

Rhue, R.D., and D.R. Hensel. 1983. The effect of lime on the availability of residual phosphorus and its extractability by dilute acid. *Soil Sci. Soc. Am. J.* 47:266–270.

Rich, C.I. 1968. Mineralogy of soil potassium. In: V.J. Kilmer, S.E. Younts, and N.C. Brady (eds.), *The role of potassium in agriculture.* Amer. Soc. Agron. pp. 79–108. Madison, WI.

Rich, C.I. 1972. Potassium in soil minerals. In: *Potassium in soil.* pp. 15–31. Int. Potash Institute, Berne.

Riga, A., V. Fischer, and H.J. van Praag. 1980. Fate of fertilizer nitrogen applied to winter wheat as $Na^{15}NO_3$ and $(^{15}NH_4)_2SO_4$ studied in microplots through a four-course rotation. 1. Influence of fertilizer splitting on soil and fertilizer nitrogen. *Soil Sci.* 130:88–99.

Riley, D., and S.A. Barber. 1971. Effect of ammonium and nitrate fertilization on phosphorus uptake as related to root-induced pH changes at the root-soil interface. *Soil Sci. Soc. Am. Proc.* 35:301–306.

Roberts, S., W.H. Weaver, and J.P. Phelps. 1980. Use of the nitrate soil test to predict sweet corn response to nitrogen fertilization. *Soil Sci. Soc. Am. J.* 44:306–308.

Rodgers, G.A. 1983. Effect of dicyandiamide on ammonia volatilization from urea in soil. *Fert. Res.* 4:361–367.

Rodrigues, G. 1954. Fixed ammonium in tropical soils. *J. Soil Sci.* 5:264–274.

Rolston, D.E. 1977. Measuring nitrogen loss from denitrification. *Calif. Agr.* 31:12–13.

Rolston, D.E., M. Fried, and D.A. Goldhamer. 1976. Denitrification measured directly from nitrogen and nitrous oxide gas fluxes. *Soil Sci. Soc. Am. J.* 40:259–266.

Rovira, A.D., and C.B. Davey. 1974. Biology of the rhizosphere. In: E.W. Carson (ed.), *The plant root and its environment.* pp. 153–204. University Press of Virginia, Charlottesville.

Rowell, D.L., M.W. Martin, and P.H. Nye. 1967. The measurement and mechanism of ion diffusion in soils. III. The effect of moisture content and soil-solution concentration on the self-diffusion of ions in soils. *Soil Sci.* 18:204–222.

Russell, J. 1973. *Soil conditions and plant growth.* 10th ed. p. 350. Longman, London.

Russell, W.J., and D.R. Johnson. 1975. Carbon-14 assimilate translocation in nodulated and nonnodulated soybeans. *Crop Sci.* 15:159–161.

Ryden, J.C., J.K Syers, and R.F. Harris. 1973. Phosphorus in runoff and streams. *Adv. Agron.* 25:1–45.

Sanders, F.E., and P.B. Tinker. 1973. Phosphate flow into mycorrhizal roots. *Pest. Sci.* 4:385–395.

Sauerbeck, D., and B. Johnen. 1976. Der Umsatz von Pflanzenwurzeln im Laufe der vegetationsperiode und dessen Beiträge zur Bodenatmung. *Z. Pflanzenern. Bodenk.* 139:315–328.

Savant, N.K., and S.K. DeDatta. 1982. Nitrogen transformations in wetland rice soils. *Adv. Agron.* 35:241–302.

Schachtschabel, P. 1980. Phosphatdüngung in Abhängigkeit vom Phosphatgehalt im Boden. *Z. Acker- Pflanzenbau* 149:191–205.

Schachtschabel, P., and G. Heinemann. 1964. Beziehungen zwischen P-Bindungs-art und pH-wert bei Lößböden. *Z. Pflanzenern. Düngung Bodenk.* 105:1–13.

Schäfer, P., and M. Siebold. 1972. Einfluß steigender Kaligaben auf Ertrag und Qualität des Sommerweizens 'Kolibri', ermittelt auf einem kalifixierenden Standort. *Bayer. Ldw. Jahrb.* 49:19–39.

Scharpf, H.C., and J. Wehrmann. 1975. Bedeutung des Mineralstickstoffvorrates des Bodens zu Vegetationsbeginn für die Bemessung der N-Düngung zu Winterweizen. *Landw. Forsch. SH* 32(I):100–114.

Scheffer, K., A. Schreiber, and R. Kickuth. 1980. Die sorptive Bindung von Düngerphosphaten im Boden und die phosphatmobilisierende Wirkung der Kieselsäure. i. Mitteilung: Die sorptive Bindung von Phosphat im Boden. *Arch. Acker- Pflanzenbau Bodenk.* 24:799–814.

Scherer, H.W. 1980. Dynamik und Pflanzenverfügbarkeit von nicht austausch-barem $NH_4^+$ im Boden. *Landw. Forsch. SH* 37:217–225, Kongreßband.

Scherer, H.W., and K. Mengel. 1981. Einfluß der Bodenfeuchte auf die Freisetzung von nicht austauschbarem $NH_4^+$ und dessen Aufnahme durch die Pflanze. *Mitt. Dts. Bodenk. Ges.* 32:429–438.

Scherer, H.W., and K. Mengel. 1983. Umsatz von [15]N markiertem Nitratstickstoff im Boden in Abhängigkeit von Strohdüngung und Bodenfeuchte. *Z. Pflanzenern. Bodenk.* 146:109–117.

Schmeer, H. 1983. Einfluß der Strohdüngung auf die Freisetzung von gasförmigen Sickstoffverbindungen. *Mitt. Dtsch. Bodenk. Ges.* 38:417–422.

Schön, M., E.A. Niederbudde, and A. Mahkorn. 1976. Ergebnisse eines 20 jährigen

Versuches mit Mineral- und Stallmistdüngung im Lößgebiet bei Landsberg (Lech.). *Z. Acker- Pflanzenbau* 143:27–37.

Schouwenburg, J.C., and A.C. Schuffelen. 1963. Potassium exchange behaviour of an illite. *Neth. J. Agr. Sci.* 11:13–22.

Schroeder, D. 1974. Relationship between soil potassium and the potassium nutrition of the plant. In: *Potassium research and agricultural production.* pp. 53–63. Int. Potash Institute, Berne.

Schroeder, D. 1976. Kalium im Boden und Kalium-Ernährung der Pflanze. *Kali-Briefe Fachgeb.* 1(3).

Schroeder, D., and H. Dümmler. 1966. Kalium-Nachlieferung, Kalium-Festlegung und Tonmineralbestand schleswigholsteinischer Böden. *Z. Pflanzenern. Düngung Bodenk.* 113:213–215.

Schubert, K.R., N.T. Jennings, and H.J. Evans. 1978. Hydrogen reactions of nodulated leguminous plants. *Plant Physiol.* 61:398–401.

Schüller, H. 1969. Die CAL-Methode, eine neue Methode zur Bestimmung des pflanzenverfügbaren Phosphates in Böden. *Z. Pflanzenern. Bodenk.* 123:48–63.

Schwertmann, U. 1966. Das Verhalten von Vermiculiten gegenüber Kalium, Aluminium und anderen Kationen. II. Chemische Verbindungen. *Z. Pflanzenern. Düngung Bodenk.* 115:200–209.

Schwertmann, U., and E. Schieck. 1980. Das Verhalten von Phosphat in eisenoxidreichen Kalkgleyen der Münchener Schotterebene. *Z. Pflanzenern. Bodenk.* 143:391–401.

Scott, A.D., and S.J. Smith. 1966. Susceptibility of interlayer potassium in micas to exchange with sodium. *Clays Clay Min. Proc. 14th Nat. Conf.* pp. 69–81.

Shiga, H., and W. Ventura. 1976. Nitrogen supplying ability of paddy soils under field conditions in the Philippines. *Soil Sci. Plant Nutr.* 22(4):387–399.

Sibbesen, E. 1978. An investigation of the anion-exchange resin method for soil phosphate extraction. *Plant and Soil* 50:305–321.

Sibbesen, E. 1983. Phosphate soil tests and their suitability to assess the phosphate status of soil. *J. Sci. Food Agr.* 34:1368–1374.

Silberbush, M., and S.A. Barber. 1983a. Prediction of phosphorus and potassium uptake by soybeans with a mechanistic-mathematical model. *Soil Sci. Soc. Am. J.* 47:262–265.

Silberbush, M., and S.A. Barber. 1983b. Sensitivity analysis of parameters used in simulating potassium uptake with a mechanistic–mathematical model. *Agron. J.* 75:851–854.

Silberbush, M., and S.A. Barber. 1983c. Sensitivity of simulated phosphorus uptake to parameters used by a mechanistic-mathematical model. *Plant and Soil* 74:93–100.

Sims, J.T., and B.G. Ellis. 1983. Adsorption and availability of phosphorus following the application of limestone to an acid, aluminous soil. *Soil Sci. Soc. Am. J.* 47:888–893.

Sims, J.R., and G.D. Jackson. 1971. Rapid analysis of soil nitrate with chromotropic acid. *Soil Sci. Soc. Am. Proc.* 35:603–606.

Singh, B., and S.P.S. Brar. 1977. Dynamics of native and applied potassium in maize–wheat rotation. *Potash Review, Subj. 9, Cereal crops 35th suite*, No. 6.

Singh, P.K. 1979. The use of *Azolla* in rice production in India. In: *Nitrogen and rice.* pp. 407–418. Int. Rice Res. Inst., Los Baños, Philippines.

Sluijsmans, C.M.J., and G.J. Kolenbrander. 1977. The significance of animal manure as a source of nitrogen in soils. In: *Proc. Int. seminar on soil environment and fertility management in intensive agriculture.* pp. 403–411. Tokyo.

Smika, D.E., H.J. Haas, and J.F. Power. 1965. Effect of moisture and nitrogen fertilizer on growth and water use by native grass. *Agron. J.* 57:483–486.

Smiley, R.W. 1974. Rhizosphere pH as influenced by plants, soils and nitrogen fertilizers. *Soil Sci. Am. Proc.* 38:795–799.

Soper, R.J., and P.M. Huang. 1962. The effect of nitrate nitrogen in the soil profile on the response of barley to fertilizer nitrogen. *Can. J. Soil Sci.* 43:350–358.

Stadelmann, F.X., O.J. Furrer, S.K. Gupta, and P. Lischer. 1983. Einfluß von Bodeneigenschaften, Bodennutzung und Bodentemperatur auf die N-Mobilisierung von Kulturböden. *Z. Pflanzenern. Bodenk.* 146:228–242.

Stanford, G., and W.H. Pierre. 1946. The relation of potassium fixation to ammonium fixation. *Soil Sci. Soc. Am. Proc.* 11:155–160.

Steffens, D. 1982. Vergleichende Untersuchungen über das Kalium-Aufnahmevermögen und die Entwicklung des Wurzelsystems von *Lolium perenne* und *Trifolium pratense*. Ph. D. Thesis, FB 19, Justus Liebig-University Giessen.

Steffens, D. 1984. Wurzelstudien und Phosphataufnahme von Weidelgras und Rotklee unter Feldbedingungen. *Z. Pflanzenern. Bodenk.* 147:85–97.

Steffens, D., and K. Mengel. 1979. Das Aneignungsvermögen von *Lolium perenne* im Vergleich zu *Trifolium pratense* für Zwischenschicht-Kalium der Tonminerale. *Landw. Forsch. SH* 36:120–127.

Steffens, D., and K. Mengel. 1981. Vergleichende Untersuchungen zwischen *Lolium perenne* und *Trifolium pratense* über das Aneignungsvermögen von Kalium. *Mitt. Dtsch. Bodenk. Ges.* 32:375–386.

Stewart, W.D.P. 1967. Nitrogen-fixing plants. *Science* 158:1426–1432.

Strebel, O., W.H.M. Duynisveld, H. Grimme, M. Renger, and H. Fleige. 1983. Wasserentzug durch Wurzeln und Nitratanlieferung (Massenfluß, Diffusion) als Funktion von Bodentiefe und Zeit bei einem Zuckerrübenbestand. Mitt. Dtsch. Bodenk. Ges. 38:153–158.

Strebel, O., H. Grimme, M. Renger, and H. Fleige. 1980. A field study with nitrogen-15 of soil and fertilizer nitrate uptake and of water withdrawal by spring wheat. *Soil Sci.* 130:205–210.

Sturm, H., and K Isermann. 1978. Überlegungen zur langfristigen Ausnutzung von Mineraldünger-Phosphat auf Ackerböden. *Landw. Forsch. SH* 35:180–192, Kongreßband.

Stutte, C.A., R.T. Weiland, and A.R. Blem. 1979. Gaseous nitrogen loss from soybean foliage. *Agron. J.* 71:95–97.

Taylor, R.W., and B.G. Ellis. 1978. A mechanism of phosphate adsorption and anion exchange resin surface. *Soil Sci. Soc. Am. J.* 42:432–436.

Terman, G.L. 1979. Volatilization losses of nitrogen as ammonia from surface-applied fertilizers, organic amendments, and crop residues. *Adv. Agron.* 31:189–223.

Teske, W., and W. Matzel. 1976. Stickstoffauswaschung und Stickstoffausnutzung durch die Pflanzen in Feldlysimetern bei Anwendung von [15]N-markiertem Harnstoff. *Arch. Acker- Pflanzenbau Bodenk.* 20:489–502.

Tomar, J.S., and R.J. Soper. 1981a. An incubation study of nitrogen added as urea to several Manitoba soils with particular reference to immobilization of nitrogen. *Can. J. Soil Sci.* 61:1–10.

Tomar, J.S., and R.J. Soper. 1981b. Fate of tagged urea N in the field with different methods of N and organic matter placement. *Agron. J.* 73:991–996.

Trolldenier, G. 1973. Secondary effects of potassium and nitrogen nutrition of rice: change in microbial activity and iron reduction in the rhizosphere. *Plant and Soil* 38(2):267–279.

van der Paauw, F. 1958. Relations between the potash requirements of crops and meteorological conditions. *Plant and Soil* 9:254–268.

van der Paauw, F. 1965. Factors controlling the efficiency of rock phosphate for potatoes and rye on humic sandy soils. *Plant and Soil* 22:81–98.

van der Paauw, F. 1971. An effective water extraction method for the determination of plant-available soil phosphorus. *Plant and Soil* 34:467–481.

van Praag, H.J., V. Fischer, and A. Riga. 1980. Fate of fertilizer nitrogen applied to winter wheat as $Na^{15}NO_3$ and $(^{15}NH_4)_2SO_4$ studied in microplots through a four-course rotation. 2. Fixed ammonium turnover and nitrogen reversion. *Soil Sci.* 130:100–105.

von Braunschweig, L.C., and K. Mengel. 1972. Der Einfluß verschiedener den Kaliumzustand des Bodens charakterisierender Parameter auf den Kornertrag von Hafer. *Landw. Forsch. SH* 26(I):65–72.

von Reichenbach, H. 1972. Factors of mica transformation. In: *Potassium in soil.* pp. 33–42. Int. Potash Institute, Berne.

Wada, H., S. Panichsakpatana, M. Kimura, and Y. Takai. 1978. Nitrogen fixation in paddy soils. I. Factors affecting $N_2$ fixation. *Soil Sci. Plant Nutr.* 24:357–365.

Walker, T.W., and J.K. Syers. 1976. The fate of phosphorus during pedogenesis. *Geoderma* 15:1–19.

Wanasuria, S., S.K. DeDatta, and K. Mengel. 1981. Rice yield in relation to electroultrafiltration extractable soil potassium. *Plant and Soil* 59:23–31.

Waring, S.A., and J.M. Bremner. 1964. Ammonium production in soil under waterlogged conditions as an index of nitrogen availability. *Nature* 201:951–952.

Watanabe, I., N.S. Berja, and D.C. Rosario. 1980. Growth of *Azolla* in paddy fields as affected by phosphorus fertilizer. *Soil Sci. Plant Nutr.* 26:301–307.

Watanabe, I., C.R. Espinas, N.S. Berja, and B.V. Alimagno. 1977. Utilization of the *Azolla–Anabaena* complex as a nitrogen fertilizer for rice. *IRRI Res. Paper Ser.* 11:3–14.

Webster, C.P. and R.J. Dowdell, 1982. Nitrous oxide emission from permanent grass swards. *J. Sci. Food Agr.* 33:227–230.

Wehrmann, J., and H.C. Scharpf. 1979. Der Mineralstoffgehalt des Bodens als Maßstab für den Stickstoffdüngerbedarf ($N_{min}$=Methode). *Plant and Soil* 52:109–126.

Weller, F. 1983. Stickstoffumsatz in einigen obstbaulich genutzten Böden Südwestdeutschlands. *Z. Pflanzenern. Bodenk.* 146:261–270.

Werner, D. 1980. Stickstoff ($N_2$)-Fixierung und Produktions-biologie. *Angew. Bot.* 54:67–75.

Werner, W. 1969. Kennzeichnung des pflanzenverfügbaren Phosphats nach mehrjähriger Düngung mit verschiedenen Phosphaten. *Z. Pflanzenern. Bodenk.* 122:19–32.

Wiklicky, L., K. Nemeth, and H. Recke. 1983. Beurteilung des Stickstoff-Düngebedarfs für die Zuckerrübe mittels EUF. *Symposium nitrogen and sugar-beet.* pp. 533–543. Int. Inst. Sugar-beet Res., Brussels.

Wild, A., D.L. Rowell, and M.A. Ogunfowora. 1969. The activity ratio as a measure of the intensity factor in potassium supply to plants. *Soil Sci.* 108:432–439.

Winner, C., I. Feyerabend, and A. von Müller. 1976. Untersuchungen über den Gehalt an Nitratstickstoff in einem Bodenprofil und dessen Entzug durch Zuckerrüben. *Zucker* 29:477–484.

Woodruff, C.M. 1955. Ionic equilibria between clay and dilute salt solutions. *Soil Sci. Soc. Am. Proc.* 19:36–40.

Young, R.A., J.L. Ozbun, A. Bauer, and E.H. Vasey. 1967. Yield response of spring wheat and barley to nitrogen fertilizer in relation to soil and climatic factors. *Soil Sci. Soc. Am. Proc.* 31:407–410.

# Microorganisms and Soil Aggregate Stability

J.M. Lynch* and Elaine Bragg†

## I. Introduction

A soil aggregate has been defined as "a naturally occurring cluster or group of soil particles in which the forces holding the particles together are much stronger than the forces between adjacent aggregates" (Martin *et al.*, 1955). The terms *soil structure* and *soil aggregation* are often used

---

*Glasshouse Crop Research Institute, Worthing Road, Littlehampton, West Sussex, BN17 6LP, England.
†AFRC Letcombe Laboratory, Wantage, Oxon, OX12 9JT, England.

© 1985 Springer-Verlag New York, Inc.
Advances in Soil Science, Volume 2.

synonymously, but soil aggregates are the basic units of soil structure, rather than the whole. Soil aggregates are formed mainly by physical forces while stabilization is effected by several factors including organic materials, iron and aluminum oxides, and clays. Sequi (1978) considered that the term "aggregation" should be used only when organic binding agents are involved, but this definition seems too narrow to us. The two processes of aggregate formation and stabilization can be concurrent in the soil, and the various stabilizing agents may act in conjunction with each other.

The size, arrangement, and stability of soil aggregates have a wide influence on soil physical properties and crop growth. The stabilization of aggregates, in which microorganisms are believed to play a major role, has received the greatest attention and has been studied for many years. Not surprisingly, a wide variety of methods has been devised to assess aggregate stability, but all of the methods are empirical and often difficult to relate to field situations. Dry aggregate stability is important in arid soils, but in most cases the stability of aggregates when they are wet is more relevant. Water, either directly as rainfall or as surface runoff, is the main agent of aggregate breakdown, so, in discussing stable soil aggregation, most workers are referring to water stable aggregation.

Methods of assessing aggregate stability range from subjecting aggregates to controlled bombardment by water drops to simulate rainfall (McCalla, 1946), to variations of Yoder's (1936) wet-sieving method where aggregates are submerged in water and the size distribution of the aggregates that remain intact is measured. Rather than assessing the size distribution of the stable aggregates, others (e.g., Gilmour et al., 1948) have determined the amount of fine material slaked from the aggregates after mechanical disruption. The stability of aggregates can also be determined by subjecting them to an internal swelling pressure, such as that resulting from successively lower concentrations of NaCl as used by Emerson (1954). Recently it has been found useful to employ volcanic ash in soil aggregation studies (Lynch and Elliott, 1983). The eruption of Mount St. Helens covered much of the Palouse soils of the Pacific Northwest of the U.S. with ash that had a particle size distribution similar to the silt loam soils of the region, the soils having been of both alluvial and volcanic origin. The advantage of this model system is that it has no interfering humic material and no inherent structure. Thus, the creation of structure by microorganisms and their products can be more precisely evaluated; such studies must, however, eventually be compared with soils *per se*.

There have been several reviews on soil aggregation and stabilization. Even in 1955 Martin et al. cited more than a hundred references on the topics, and since then several others have reviewed the literature on the subject, the most extensive being that of Harris et al. (1966a).

Today synthetic and natural soil conditioners can be used to improve soil aggregate stability (Page, 1983), if not on a field scale then at least in small-scale horticultural uses. Since there exists the prospect of the use of microorganisms for the production of aggregate stabilizing agents, either in culture or in the controlled decomposition of crop residues to be added back to the soil, interest in the role of microorganisms and their products in aggregate stabilization has recently been revived.

## II. Why Is Stable Aggregation Necessary?

Soil aggregates, composed of primary particles and binding agents, are the basic units of soil structure. The main effects of poor soil structure on crop yield are through its effects on surface instability, soil compaction, and the presence of anaerobic zones in the soil (Batey, 1974). The importance of soil aggregation in crop production lies in its indirect effect on water and air relationships in the soil. The size, shape, and stability of soil aggregates control the pore size distribution, which in turn affects many soil physical properties.

A productive agricultural soil must have a wide range of pore sizes. Coarse pores allow rapid infiltration of water and allow roots of young plants to grow through them. Pores larger than 30 to 60 μm allow water to drain away under the influence of gravity while water available to plants is held in pores of 0.1 μm to 15 μm. Gaseous exchange between the soil and the atmosphere is mediated by the coarser pores (>60 μm) from which water drains under the influence of gravity. A range of pore sizes is necessary for root growth and extension because although roots can expand narrow pores they cannot grow into very compact soil.

Kilbertus (1980) investigated the pores of a Rendzina, a Chernozem, and an Acid Brown soil and found the mean diameter of the pores to be about 2 μm. The size of the bacteria was variable but always less than 1 μm, and the ratio of the mean diameter of bacteria to the mean diameter of pores was constant at 1:3. Some pores were completely closed whereas others were open, and he discussed this in relation to gas diffusion.

Although it may be more relevant to describe aggregation in terms of the stable pores in the soil, it is much more straightforward to assess the size and stability of the aggregates that surround the pores. If the aggregates are unstable, they can slake when wetted, and detached microaggregates, or even clay particles, may be carried into pores making them narrower or discontinuous. Slaking occurs when aggregates are not stable enough to withstand the pressures resulting from entrapped air that build up when aggregates are wetted rapidly.

The stability of surface aggregates is most important because aggregates below the surface are protected from rapid wetting by those above. Unstable surface aggregates can lead to the formation of crusts that inhibit the movement of water and air into the soil. The rate of oxygen diffusion in a soil was reduced by 50% within 24 hrs of the formation of a wet surface crust (Rathore et al., 1982). This may be particularly critical for seed germination (Lynch and Pryn, 1977) since, by contrast with roots, there are no mechanisms by which $O_2$ can be transported to the seed. A restriction of oxygen diffusion can also lead to the formation of anaerobic zones, altering the chemical and microbiological status of the soil. When dry, a surface crust can pose a considerable mechanical barrier to seedling emergence.

Under modern agricultural systems, the instability of soil aggregates can be a major limiting factor to the production of arable crops. Aggregate stability changes in response to cropping sequences and organic matter levels in the soil, decreasing on cultivation of a virgin soil and increasing under long-term pasture. The organic matter level in soil tends to reach an equilibrium level (Johnston, 1982), depending on the farming system, climate, and soil type, and is affected only slowly by annual additions of plant materials or manures. In North America, unlike Britain, many soils have not yet reached such an equilibrium. Cultivation, especially if a fallow is included in the rotation, decreases the amount of organic matter in the soil, partly by exposing it to oxidation. In a sandy loam soil in England, 100 yrs of continuous cereals reduced the organic carbon content by 50% (Johnston, 1982), while annual additions of farmyard manure over 150 yrs increased the organic carbon content from 1% to 3%. In plots established since 1912 on a Dark Brown Chernozemic soil in Canada, Dormaar and Pittman (1980) found that cultivated soils contained 47% less C, 46% less N, 53% fewer polysaccharides, 100% more solvent-extractable C, and 49% more resin-extractable C and had a slightly higher pH than the proximate grassland soil.

Good, stable aggregation can be damaged by the intensification of arable cropping. The replacement of traditional cropping systems by mono-cultures can result in a reduction in the stability of the soil aggregates, attributed mainly to the removal of grass leys from the cropping sequence. The use of heavier machinery with more frequent working of the soil, especially if it is wet, can damage the soil structure by increasing its bulk density. Under monocultures of barley (*Hordeum vulgare*) in Denmark, Hansen (1982) found that the pore space and air content of the soil were reduced by increases in compaction caused by traffic on wet soils. However, on some soils, modern systems of reduced cultivation may improve the state of aggregation of the soil. Shallow tine cultivation and direct-drilling of winter cereals increased aggregate stability in soils with a high clay content (Douglas and Goss, 1982), although a silty soil gave the opposite result, but Hamblin (1980) found no consistent differences in the stability of aggregates from direct-drilled soils or ploughed soils sown with wheat (*Triticum aestivum*) or barley. It should be noted that in these studies straw from the preceding crop was burned prior to drilling; this is normal in Britain and contrasts with North American practice.

In their efforts to keep up with world food demand, many farmers have adopted policies that are leading to excessive rates of soil erosion (Brown, 1981). Pressure to increase food production can mean that land previously thought to be unsuitable for cultivation is brought into use. The soil may have been stable under its climax vegetation, but, when brought into crop production, the organic matter content may decrease, one result being that aggregates become less stable. Yield on eroded soil is often lower and more variable. In some cases soil erosion on a dramatic scale may result, but more insidious is the slow removal of the topsoil that is occurring in many parts of the world. In the U.S., water erosion alone annually removes some 2 billion tons of topsoil—just over a billion tons more than is formed each year (Brown, 1981).

Given the importance of soil aggregates in determining the water and air relationships in the soil and the potentially disastrous consequences of mismanagement, the processes involved in the formation of soil aggregates and their stabilization have been the subject of a great many studies over the past 50 yrs. The formation of aggregates is attributed to mainly physical forces such as wetting and drying, freezing and thawing, and the compressive and drying action of roots. Once the primary particles have been brought into close proximity to each other, various factors appear to be responsible for binding them together and thus stabilizing the aggregate.

The main agencies of stabilization are organic materials. These include the products of decomposition of plant, animal, and microbial remains; the microorganisms themselves; and the products of microbial synthesis. Various inorganic cementing agents provide stabilization in certain soils [e.g., iron oxides may be the predominant binding agent in lateritic soils (Lutz, 1936)], but in the majority of agricultural soils, organic binding agents are of the greatest importance.

## III. How are Microorganisms Involved?

The improvement of soil physical conditions brought about by the addition of organic matter has been known for a long time. For centuries organic materials, in the form of farmyard manure, crop residues, and green manures, have been added to soil, and even in the time of Cato (239–149 B.C.) (cited by Gati, 1982), the ploughing in of lupins, vetch, and beans was advised. Evidence for the involvement of microorganisms in soil aggregate stabilization comes from early work on the effects of adding organic materials to soil. Organic matter additions have no effect unless microorganisms are present (Martin and Waksman, 1940; McCalla, 1945). In the presence of microorganisms, increases in aggregation occurred over one week of incubation with little increase with longer time (McCalla et al., 1957). Most organic matter additions have little immediate effect, but changes in stability after the addition of substrates have been detected after 1 hr by Skinner (1979), suggesting a physical cause in some cases.

The increase in stability is directly related to the ease of microbial decomposition of the added organic material (Martin and Waksman, 1940, 1941), additions of alfalfa being more effective than peat. The duration of the aggregating effect and the time taken to reach the maximum depend on the nature of the substrate (Martin and Waksman, 1941). Less easily decomposed material such as cereal straw or cellulose produced less stable aggregates in the same period of time than an easily decomposed substrate such as sucrose (McCalla *et al.*, 1957) or glucose (Griffiths and Jones, 1965), probably because the latter are more readily available to micro-organisms. The ability of some fungi to stabilize non-amended aggregates as well as sucrose-amended aggregates led Harris *et al.* (1966b) to deduce that they were able to form binding agents from soil organic matter as well as from sucrose. Martin (1942) found that composting organic materials decreased the percentage of stable aggregates formed when they were added to the soil. After composting, much of the readily available material may have been utilized, leaving less easily decomposed material that is less effective in aggregate stabilization. Therefore, it may be best to add fresh material that is readily available to microorganisms. The carbon:nitrogen ratio of the substrate affects its aggregating effectiveness. The percentage of water stable aggregates increased when ground wheat straw was incubated in soil, but there was a decrease in aggregation when nitrogen was added with the straw (Acton *et al.*, 1963). Skinner (1979) found that when peptone was added to natural soil aggregates and incubated there was usually a decrease in aggregation, presumably due to the extra available nitrogen promoting greater microbial utilization of carbon-rich binding agents, some of which may already have been present in the soil.

The environmental variables that affect soil aggregate stability after organic matter addition are those that affect the activity of the microbial population, although it is difficult to make generalizations. pH conditions unfavorable to microbial activity restricted aggregate stabilization (Aldrich, 1948), but Swaby (1949) found that adjustment of pH in the range of 4.75 to 7.5 had no appreciable effect on aggregation. pH may affect the degradation of microbially produced stabilizing agents (Martin and Aldrich, 1955).

As the temperature of incubation was increased, the time required to reach the maximum aggregation decreased (Martin and Craggs, 1946). Martin and Craggs (1946) also found that more resistant organic materials were relatively more effective aggregate stabilizers at higher temperatures. Using Peorian loess, McCalla *et al.* (1957) found no effect on aggregation of temperatures between 20 and 28°C, but Harris *et al.* (1966b) found that maximum stability of sucrose-amended artificial aggregates containing the indigenous population was maintained for a shorter time at 35°C compared with 15 or 20°C. The rate of stabilization but not the final level was increased by the increase in temperature, and when the indigenous population was present, the numbers of organisms were increased more quickly at higher temperatures. At different temperatures, different populations may exist. Martin and Craggs (1946) postulated that a greater quantity of better quality aggregating materials may be produced at lower temperatures, but higher temperatures may also favor more rapid degradation of the binding agents.

Conflicting results have been obtained on the effects of moisture content and aeration, indicating the balance between the formation and degradation of stable aggregates. Under waterlogged conditions, organic materials were less effective in improving aggregation than when the soil was only 25 to 75% saturated (Martin and Craggs, 1946). In contrast, McCalla *et al.* (1957) found that increases in soil moisture content from 25 to 30% had no great effect on aggregation by three fungi. Stability under aerobic incubation was usually higher than anaerobic (Skinner, 1979), but Harris *et al.* (1963) found that anaerobic incubation of sucrose-amended artificial aggregates gave stable aggregates over a long time, leading them to suggest that anaerobic bacteria were unable to decompose the binding substances they produced, whereas under aerobic conditions the microorganisms were able to utilize both sucrose and the stabilizing agents they produced. Aggregates that had been stable under anaerobic conditions broke down on exposure to air, perhaps because of breakdown of the stabilizing agents formed previously.

By observing thin sections of aggregates, Griffiths and Jones (1965) noted that there was a general relationship between the change in stability and colonization by fungi, yeasts, and actinomycetes. However, in field soils only poor correlations have been found between microbial numbers in the soil and its state of aggregation, probably because some groups or species are more effective than others. Meyers and McCalla (1941) did find a significant positive correlation between bacterial numbers and degree of aggregation after 8 and 18 days incubation. In non-sterile soil there was no correlation between microbial numbers and aggregation during the early stages of incubation (Swaby, 1949). Because there was no relationship between bacterial numbers and aggregate stability, Harris *et al.* (1966b) deduced that there was little effect of the cells *per se*. Meyers and McCalla (1941) found, in a second experiment, that maximum aggregation lagged behind the maximum of bacterial numbers, suggesting that microbial products were more important than microbial numbers. However, it is difficult to interpret microbial counts, and a more meaningful measure would be of microbial biomass in relation to aggregation (Lynch, 1981a). Our recent studies have demonstrated highly significant positive correlations between the number of cells added to a soil and the improvement in stability. In this context cells added are meaningful because, unlike cells counted from soils, they are directly proportional to biomass. Both the whole cells and the extracted polymer are effective, but the greatest effect is obtained with whole cells.

The net effect of the soil microflora on soil aggregation is almost impossible to predict (McCalla *et al.*, 1957), but various microbial groups have been found to be responsible for aggregate stabilization, varying in their effectiveness (McCalla, 1946; Harris *et al.*, 1964). Specific examples are reviewed by Harris *et al.* (1966a), but in general fungi have been found to be the most effective, with large variations between species of bacteria (Peele, 1940). Fungi vary in their effectiveness (Gilmour *et al.*, 1948), depending on the organic amendment added, with some species being better able to decompose complex organic materials. Fast-growing fungi with woolly hyphae may be the most effective (Swaby, 1949), although Harris *et al.* (1966b) found no relationship between mycelial form and aggregation. McCalla (1946) compared the effects of different taxonomic groups on aggregate stability. He found that actinomycetes were poorer aggregators than fungi, and that although all bacteria were less effective than fungi or actinomycetes, some were better than others.

In summary, the addition of organic materials to soil can bring about an increase in the stability of the soil aggregates. This effect is due to microbial activity and is affected by environmental factors. The ability of organisms to mediate in aggregate stabilization seems to be widespread although some may be more effective than others.

There are two major ways in which microorganisms may be involved:

1. Some organisms may be able to mechanically bind soil particles together.
2. Others may produce effective binding agents either by synthesis or through the decomposition of organic materials. These products may remain in close contact with the cells or become part of the pool of soil organic matter and subject to decomposition.

## A. Mechanical Binding of Soil Particles

In considering aggregate stabilization by microorganisms, it is rarely possible to separate mechanical effects from the effects of products closely associated with the cells. From size considerations it seems likely that fungi are involved in binding together larger soil particles and that bacteria may be involved in microaggregate stabilization of clay particles. In an earlier review (Fletcher *et al.*, 1980) this was depicted schematically. However, recent evidence using electron microscopy (Campbell, 1983) indicates that clay particles adhere to the walls of fungal hyphae. This is allowed for in Figure 1 where only a detail of the soil aggregate is shown. The scheme is supported with evidence obtained by prizing open soil aggregates and viewing them by scanning electron microscopy; bacteria (Figure 2) and fungal hyphae (Figure 3) are evident. Bacteria can also be seen by preparing thin sections of aggregates and viewing with the transmission electron microscope (Figure 4).

The role of fungi may be considered as both aggregate forming and aggregate stabilizing. By ramifying through the soil fungal hyphae may

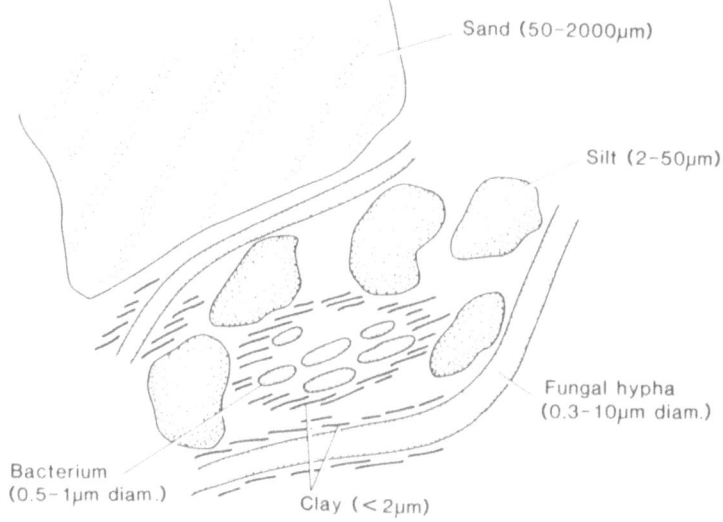

**Figure 1.** Schematic representation of microorganisms binding soil particles.

bring soil particles together and force their contact with binding agents. Filamentous fungi differed markedly in their stabilizing abilities, the most effective being vigorous growers that produced woolly hyphae on both agar and soil, e.g., *Mucor sp.* (Swaby, 1949). The degree of water stable

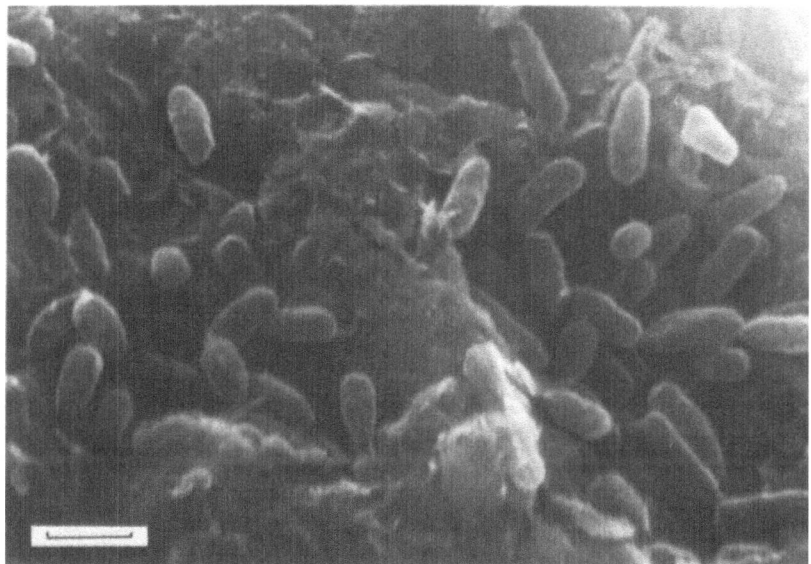

**Figure 2.** Bacteria within a soil aggregate (bar marker 1μm).

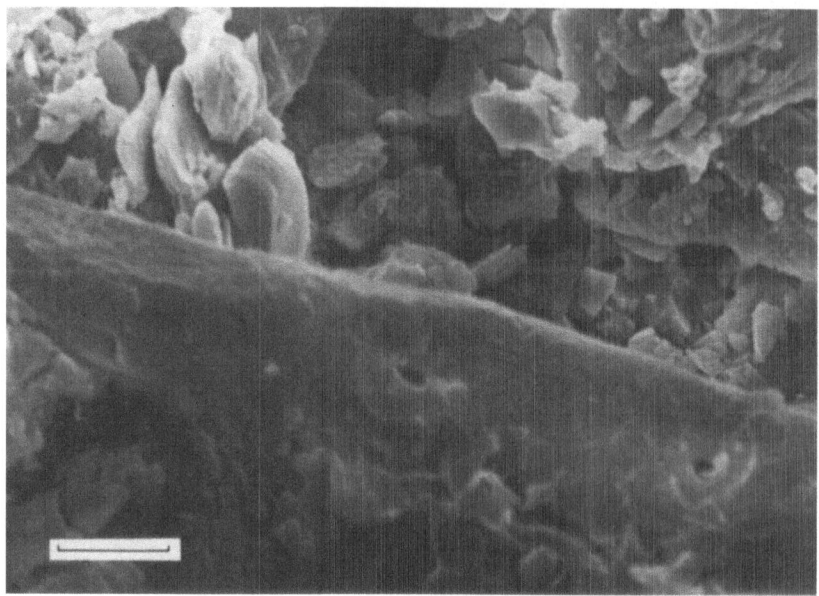

**Figure 3.** Fungal hyphae within a soil aggregate (bar marker 10μm).

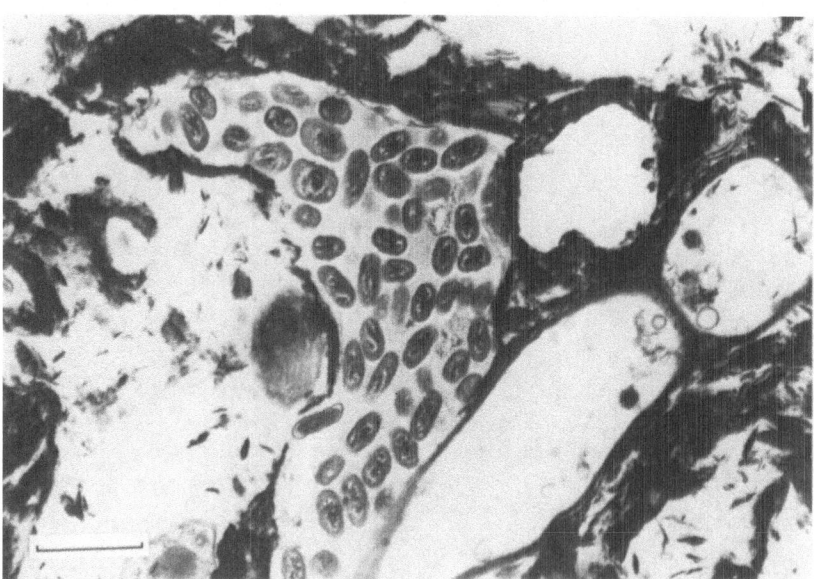

**Figure 4.** Bacteria surrounded by clay particles in the closed pore of a soil aggregate (bar marker 1μm). Photograph courtesy of Dr. G. Kilbertus, University of Nancy.

aggregation correlated well with hyphal length. The effect of fungi on stability may be purely physical but, where hyphae have been observed to bind soil particles together, grains or small aggregates also seem to be attached to the hyphae. Because the aggregate stabilizing effect of extracts of microbial cultures is often less than that attributed to the products of microbial decomposition of organic matter, Went and Stark (1968) suggested that aggregation in field soils may, at least in part, be due to mechanical binding by fungal mycelium. Martin (1945) estimated that at least 50% of the stabilizing effect of fungi was due to mechanical binding while the rest was due to products of decomposition or synthesis, although the decrease in stabilizing efficiency in the absence of cells could be due in their experiments to changes in the distribution of soil binding agents.

Several investigators have observed enmeshment of soil or sand particles by hyphae (Bond and Harris, 1964; Gray, 1967; Aspiras et al., 1971b), and Watson and Stojanovic (1965) concluded from microscopic examination that the effects of fungi appeared to be largely physical. Harris et al. (1966b) found that fungi had little effect on stability until macroscopic mycelium appeared. The filaments observed by Bond (1959) were strong enough to withstand sieving, and later examination of several sandy soils showed that well-structured soils always had hyphae present while poorly structured soils were lacking in hyphae, with little seasonal variation in old pasture soils (Bond and Harris, 1964). Discussing the role of hyphae as temporary binding agents of aggregates <250 μm, Tisdall and Oades (1982) suggested that saprophytic fungi, which are often absent on dilution plates and are therefore seldom studied, may play an important part in aggregation since they tend to persist longer in soils than other fungi (Martin et al. 1959). However, this suggestion is difficult to substantiate becasue there is a lack of knowledge on the longevity of the various groups of soil organisms.

Although part of the effect of fungi on stabilization may be mechanical, the attachment of particles to hyphae has often been observed (Bond, 1959; Aspiras et al., 1971b) and binding agents detected (Clough and Sutton, 1978). Fungi may be effective stabilizers because the spread of hyphae between aggregates and into large pores distributes their associated binding agents through the soil. Their importance in aggregate stabilization may be more indirect, in that they can serve as substrates for other microorganisms.

Other filamentous organisms have been implicated in stable aggregate formation. Watson and Stojanovic (1965) found that aggregates produced by streptomycetes were larger than those of bacteria and were penetrated by filaments. Aspiras et al. (1971b) observed soil particles adhering to streptomycete hyphae, indicating the presence of soil binding agents on the hyphae. Bond and Harris (1964) found that lichens and algae formed surface crusts in sand in a variety of habitats. In low rainfall areas of Australia they found that crusts of sand were interwoven with algal filaments that had bacteria and fungi associated with them. Some forms of blue-green algae in which filaments are surrounded by mucilaginous sheaths would be well adapted to binding sand.

It is questionable whether bacteria can generally be said to have a direct mechanical effect on binding of soil particles, even of fine clays. The increase in aggregate stability after the introduction of organic materials generally lags behind the increase in microbial numbers, suggesting superficially that they are not directly responsible for stabilization. The ability of bacteria to become associated with soil and other surfaces is known. The work of Stotzky and his co-workers has shown that bacteria can adsorb to the surface of pure clays (Santoro and Stotzky, 1968; Stotzky and Bystricky, 1969). Adsorption of *Rhizobium* sp. to silt particles has been observed by Fehrmann and Weaver (1978), who concluded that the adsorption of bacteria and the aggregation of the silt particles was due to the presence of polysaccharide material produced by the bacteria. Since it is very difficult to separate cells from their surface polymers, it is likely that material associated with the cell walls mediates in their interaction with clay particles.

### B. Influence of Microbial Products

Microbial decomposition of plant tissue in the soil yields two major types of organic residue: (1) resistant compounds, e.g., fats, lignins, and waxes; and (2) newer compounds that may either be products of the decomposition of plant and other tissues or may be freshly synthesized by soil micro-organisms.

The end product of decomposition is humus, a dark-colored, hetero-geneous colloidal mixture. Soil chemists have separated humus into insoluble humin and soluble humic colloids, although different workers use different terms. The humic colloids can be further separated into fulvic acids and humic acids. The composition and structures of humic colloids have been discussed by Hayes (1980). Briefly, the humic colloids include polysaccharides, proteins, and many substances of uncertain composition but which contain a large number of aromatic rings.

Aggregate stability was decreased by removal by hydrogen peroxide of the soil organic matter associated with the clay fraction (Robinson and Page, 1950), and several attempts have been made to determine the fraction of the soil organic matter associated with aggregate stabilization. Positive correlations have been obtained between the level of carbohydrate in the soil and its state of aggregation (Watson and Stojanovic, 1965), suggesting that carbohydrate may have contributed to aggregation, in this case by mixed cultures. The content of microbial gum in soils was the most important factor influencing soil stability for four soil types (Chesters *et al.*, 1957), while Acton *et al.* (1963) found that when two soils were incubated with ground wheat straw their microbial gum content correlated well with the number of aggregates >0.5 mm but not with aggregates >0.1 mm. When the organic matter was extracted and chemically fractionated, the polysaccharide content of the precipated gum fraction of the fulvic acid also correlated with the aggregating effect. Rennie *et al.* (1954) also obtained good correlations between the polysaccharide content of two silt loam soils and their state of aggregation. By inoculating single species of microorganisms into sucrose-amended aggregates and incubating them before subjecting them to a variety of chemical treatments, Aspiras *et al.* (1971a) found that polysaccharides were the main factor responsible for aggregate stabilization of aggregates by bacteria. However, a single alkali extraction, as used by Chesters *et al.* (1957), may not remove all of the polysaccharide present in the soil; indeed, it is likely to remove only the most readily extractable polysaccharide (Mehta *et al.*, 1960). Being easily removed, it is likely to be least adsorbed to the soil particles and may participate least in aggregate stabilization.

Mehta *et al.* (1960) put forward the suggestion that if polysaccharides were responsible for aggregate stability then a chemical that breaks down polysaccharides should reduce the stability of aggregates. Sodium periodate oxidizes sugars by cleaving the bonds between adjacent carbon atoms carrying −OH groups (Bobbit, 1956). The partially oxidized polymers (polysaccharides and polyuronides) are then unstable in alkaline conditions, so polymers cleaved in this way can no longer act as bridges between soil particles. Mehta *et al.* (1960) treated aggregates of a Swiss Braunerde soil with sodium periodate, but the stability of natural aggregates was not affected by treatment. However, the presence of $CaCO_3$ in the soil they used may have interfered with the reaction, making the aggregates stable to periodate. Removal of $CaCO_3$ from one Houston soil made its crumbs more susceptible to periodate oxidation (Clapp and Emerson, 1965).

Greenland *et al.* (1961) found that stability resulting from periodate-sensitive materials was greatest in soils with low organic matter contents and postulated that the material responsible for periodate-resistant aggregate stability in long-term pasture soils may be fungal hyphae since these are common in those soils but not in young pasture soils. The same authors later found that in a Red-Brown Earth soil, aggregate stability resulting from periodate-insensitive materials increased with the number of years a soil had been under pasture, while soils that had just been cultivated owed most of their stability to periodate-sensitive materials (Greenland *et al.*, 1962).

Because of the conflicts in previous work on the effects of periodate oxidation on aggregate stability, Clapp and Emerson (1965) examined the effects of sodium chloride, borate, pyrophosphate, and periodate on aggregate stability. Crumbs from grassland soil were only slightly affected by periodate, but if they were first extracted with pyrophosphate, they broke down completely. Since pyrophosphate treatment alone did not cause aggregate breakdown, they postulated that two types of binding agent were responsible for their stabilization. The stability of cultivated crumbs was destroyed completely without a preliminary extraction by pyrophosphate. However, because periodate treatment may react with the *cis*-hydroxyl groups of molecules other than polysaccharides and break them down, all that can really be said about stabilizing materials that are sensitive to periodate is that they contain these groups. Some recent findings of Cheshire *et al.* (1983) suggest that in earlier studies, oxidation of polysaccharide by periodate was by no means complete under the conditions used. Although the majority of microbial polysaccharides may have been oxidized, carbohydrate still persists in the soil and may be present in plant fragments that could contribute to aggregation.

Apart from polysaccharides, other materials produced by microorganisms can have an effect on stability. For example, by using various chemical extractants, Aspiras *et al.* (1971a) found that, in addition to polysaccharides, humic- and ligninlike substances and waxy or resinous materials were in part responsible for stabilization by fungi. However, the use of chemical extractants to remove humic materials has been criticized because they may also remove Fe, Al, and Ca ions that may also be active in aggregate stabilization (Giovaninni and Sequi, 1976), and small, but perhaps significant, amounts of polysaccharide. Geoghegan and Armitage (1949) extracted the mycelium of *Penicillium notatum* and *Aspergillus niger* and studied the effect of the extracts on aggregate stability. Spraying the extract onto the aggregates had a much greater effect than incorporating it into them. The extracts also made the aggregates water repellant. Geoghegan (1950) investigated the aggregating effect of lipoidal substances produced by fungi. These substances waterproofed aggregates and again were more effective when sprayed onto aggregates than when incorporated into them.

The effect of these hydrophobic substances on aggregate stability has been studied by few investigators, yet soil wettability is an important factor in soil erosion, erosion being decreased when hydrophobicity is lowered (Giovaninni *et al.*, 1983). Using benzene as an extractant, they found that the hydrophobicity of a naturally water-repellant soil was reduced, as was aggregate stability. The polar substances extracted therefore seemed to increase water stability, but by also increasing the hydrophobicity of the soil they would increase the runoff of water from the surface and perhaps increase erosion.

However, the bulk of the available evidence points to carbohydrates, specifically polysaccharides, as being a major factor in aggregate stabilization, at least in some soils, and especially in those in cultivation. The polysaccharides isolated from soil are generally very complex (Martin, 1971) and are a mixture of large, usually linear, flexible polymers. Analysis of the constituent sugars present in soil polysaccharides gives indirect evidence of their origins. Few plant polysaccharides exist in the soil at any time because of their rapid decomposition (Swincer *et al.*, 1969). The relatively low contents of glucose and xylose units, which are predominantly plant polysaccharides, and the relatively high contents of mannose, rhamnose, and hexosamines, indicate that the majority are of microbial origin. More direct evidence comes from work such as that by Keefer and Mortensen (1963) using $^{14}$C-labeled substrates, which indicates that soil microorganisms are capable of synthesizing all of the kinds of neutral sugars found in soil. In inoculated soil, Sparling et al. (1981) found an increase in the amounts of mannose, galactose, and rhamnose in the soil, indicating that the polysaccharides synthesized in soil were similar to those in pure culture. However, quantitative work by Cheshire et al. (1969) suggests that the various sugars found in soil after incubation with labeled substrate may not be produced in the same proportions as those found in native soil polysaccharides. Cheshire (1977) concluded that the hexoses and doxyhexoses in soils are derived from both plants and microorganisms, while the pentose sugars are present in residual plant polysaccharides. Particle size fractionation of soil in which $^{14}$C-labeled substrates had been incubated showed that all fractions contained sugars, with much greater amounts being present in the clay and silt fractions (Cheshire and Mundie, 1981).

The ability of soil microorganisms to produce extracellular polysaccharides in pure culture is well known (Finch et al., 1971). Both Gram-negative and -positive organisms can produce extracellular polymers (Sutherland, 1977), and although they are probably not essential to survival (Dudman, 1977), polysaccharides may have a role in virulence or serve as protective mechanisms against predation or dessication. Bacteria that could produce polysaccharides in the laboratory were found to be present in all of the soils tested by Forsyth and Webley (1949), and many other workers have isolated polysaccharide-producing bacteria from soil.

Martin (1945) found that three of the polysaccharide-producing bacteria he examined were able to produce polymer on plant residues, indicating that polysaccharide production would take place in soil. Martin (1975) concluded that microorganisms were able to utilize water-soluble materials released from the rhizosphere of plants. Most of the material recovered from the soil by leaching in the later stages of plant growth was derived from microorganisms because it was similar in composition to that from the decomposition of dead roots incubated in soil. Swincer et al. (1969) suggested that many of the polysaccharides that persist in soil are probably derived from recently dead microorganisms because hexosamines and other sugars commonly found in the soil are constituents of bacterial cell walls. However, Finch et al. (1971) considered that if appreciable amounts of soil polysaccharides were of microbial origin, then (based on calculations of the contribution of the live soil biomass to soil polysaccharide) the microbial materials must have a high degree of resistance to decomposition. The possible mechanisms by which polysaccharides may be protected from degradation are discussed later.

Early work on the ability of the products of microbial synthesis to stabilize aggregates showed that the material isolated from pure cultures of bacteria could stabilize aggregates and that relatively small amounts of some were necessary to increase stability (Martin, 1945). The addition of washed cells of *Bacillus subtilis* had little effect on aggregate stability (Geoghegan and Brian, 1946), but when cultured, *B. subtilis* produced a levan, the aggregating effect of which depended on its viscosity. As the amount of sucrose in the medium increased, the molecular weight of the polymer, as measured by viscosity, decreased, and its aggregating effect decreased (Geoghegan and Brian, 1948). Several workers have found a similar relationship between the viscosity of bacterial polysaccharides and their aggregating effect (Clapp *et al.*, 1962; Dabek-Szreniawska, 1972; Gaur and Rao, 1975; Elliott and Lynch, 1984). Gaur and Rao (1975) also found a positive relationship between the uronic acid content of bacterial polymers and their stabilizing effect.

The evidence for the involvement of polysaccharides in soil aggregate stabilization has been summarized by Burns (1977):

1. A positive correlation is found between extractable polysaccharide and aggregation.
2. Treatment of soil with sodium periodate can result in the disruption of aggregates.
3. Many soil microorganisms produce extracellular polysaccharides, at least *in vitro*.
4. Aggregation is stimulated by the addition of easily degradable carbon sources to soil.
5. Soils amended with microbial polysaccharides show improved stability.

## IV. What Are the Mechanisms Involved?

Tisdall and Oades (1982) stress that in discussing the mechanisms of aggregate stabilization it is necessary to consider the scale at which the various associations take place. They visualize aggregate stabilization at several different levels where particles of <0.2 μm are built up into larger aggregates,

$$<0.2 \ \mu m \rightarrow 0.2\text{--}2 \ \mu m \rightarrow 2\text{--}20 \ \mu m \rightarrow 20\text{--}250 \ \mu m \rightarrow >2000 \ \mu m,$$

with different agents being responsible for stabilization at each level. For example, they suggest that 2 to 20 μm aggregates are bound together by persistent organic bonds while aggregates >2000 μm are held together mainly by a network of roots and hyphae.

Having already discussed the ways in which microorganisms may be involved in aggregate stabilization, the mechanisms by which they may be effective will now be considered:

1. Polymers produced by bacteria may adsorb to soil surfaces.
2. By themselves adsorbing to soil particles, microorganisms may bind soil particles.
3. Groups of microorganisms may interact with each other or with roots to stabilize aggregates.

These three mechanisms could also be thought of as occurring at different size levels, the first two leading to the formation of microaggregates and the third leading to a higher level of organization.

### A. Binding by Organic Molecules

Polysaccharides synthesized by soil microorganisms can be divided into two groups: homopolysaccharides, which include levans and dextrans and are uncharged; and heteropolysaccharides, generally composed of repeating sugar units, often with uronic acid groups present. These polysaccharides can be linear or branched with a variety of functional groups such as hydroxyls and carboxyls. At the pH's that exist in soils, many would be expected to be negatively charged, owing to dissociation of their functional groups. Similarly, clay surfaces have a negative charge arising from isomorphous substitution within the clay lattice, although some positive charge can develop at the edges of clay platelets. Some clay minerals also have a pH-dependent negative charge.

Several reviews have described the isolation and composition of soil polysaccharides (e.g., Cheshire, 1979; Finch *et al.*, 1971). Until recently, the lack of purity of isolates obtained from soil has limited clarification of their structures, but using new techniques more accurate determinations can now be made. For example, Newman *et al.* (1980) have used nuclear magnetic resonance to characterize humic substances. Extraction and fractionation of soil polysaccharides have revealed a complex mixture of sugar units, perhaps because of deficiencies in the separation procedures, but perhaps because the soil does contain very complex molecules (Martin, 1971). During degradation, microbial polysaccharides could be built up to form new polymers, perhaps through the action of free enzymes. Burns (1983) has suggested that extracellular enzymes, or those from lysed cells, may survive for a long time in soil, possibly because of interactions with organic and inorganic soil components.

Emerson (1959) proposed a model of a soil crumb in which organic materials linked clay domains and quartz particles. The quartz–clay bonds were strengthened by organic polymers linking the quartz surface and the edge or basal surfaces of the clay. Many studies have been made of the adsorption of organic molecules to clays, and there have been several reviews on the subject, notably those of Martin (1971), Greenland (1972), Hayes (1980), and Theng (1982).

The adsorption of short-chain, low molecular weight, neutral molecules to clays can be fairly accurately predicted, but organic macromolecules have high molecular weights, long chains, and are often charged, so that they take up random conformations in solution and can adsorb to surfaces at many points. Adsorption from solution depends predominantly on the length of the organic molecule and the net charge on the polymer and the clay, both of which are affected mainly by the solution pH and the presence of cations.

There is much evidence that increasing polymer molecular weight increases the strength of adsorption (Greenland, 1965a). Comparing the adsorption of dextrans of different molecular weights, Parfitt and Greenland (1970) found that two lower molecular weight ones were not adsorbed by Na-saturated montmorillonite, whereas one with a higher molecular weight was strongly adsorbed by this but less so by Ca- and Al-saturated clay. This lack of adsorption of the lower molecular weight polymer could be due to changes in the molecular configuration of the polymer (Parfitt and Greenland, 1970). This is known to affect the interaction with water, the desorption of which from the clay surface is likely to be involved in adsorption. Changes in configuration could occur at the clay surface.

The complexity of the molecule and the nature of its functional groups are also important. Olness and Clapp (1975) found that the primary structure of the polymer could influence adsorption. The number of primary hydroxyl groups determined the length of the segment that had to be adsorbed to form a stable complex. Later, the same authors (Olness and Clapp, 1976) compared two polysaccharides, differing mainly in their ratio of primary to secondary OH groups, and concluded that differences in the intermonomer bonding in the chain, the type of branching, and substitution into linear chains all have an effect on adsorption.

The pH of the system has a large effect on adsorption through its effect on the ionization of the functional groups of the polymer. Negatively charged polymers would be expected to be repelled by the like-charged clay surfaces, but adsorption of these polymers can take place through cation bridges. Parfitt and Greenland (1970) found that at pH 6 polygalacturonic acid was excluded from the surface of Na- and Al-saturated montmorillonite clay because the functional groups of the polymer would largely be dissociated at this pH while at lower pH's adsorption may take place through polyvalent cation bridges, especially those involving Al.

The presence of different cations at the clay surface affects adsorption (Parfitt and Greenland, 1970). Multivalent cations at the clay surface can form ligands with anionic groups of polymers forming cationic bridges between clays and organic compounds. Microbial metabolites (from spent media) induced flocculation of clays homoionic to La or Al (Santoro and Stotzky, 1967).

The actual bonds that are involved have been discussed by Greenland (1970), Burns (1979), and Theng (1982) and so will be referred to only briefly here. Many types of interaction can take place. For neutral polymers, hydrogen bond formation may be the most important. Uncharged polymers may be adsorbed in the interlamellar region of clays, but negatively charged polymers are adsorbed only on surfaces or at edges (Greenland, 1965b). Cationic polymers may participate in normal ion-exchange reactions, but the interaction is more complex for negatively charged polymers, which would be expected to be repelled by the negatively charged clay surfaces. At pH values above 2 to 3, i.e., above their iso-electric point, most microbial and soil polysaccharides are negatively charged. They may interact with surfaces through polyvalent cations, forming ligands, or may take part in coulombic interactions with positively charged edges of clay particles.

Hydrogen bonding may also be important and has been suggested by several workers. Lynch *et al.* (1956) found that simple carbohydrates were adsorbed by clays in lesser quantities than more complex ones. X-ray diffraction measurements showed that carbohydrates were adsorbed between the clay plates, and data from infrared spectra suggested that H-bonding may be involved. Greenland (1956a) found that soil polysaccharide was very firmly bound to clay, and X-ray analysis showed that it was adsorbed in the interlamellar region of the clay. Using X-ray analysis to measure the basal spacing of clays complexed with sugars, Greenland (1956b) found that carboxylated sugars were less strongly adsorbed and methoxylated sugars more strongly adsorbed than unsubstituted sugars. This indicates a lack of involvement of H-bonding in unsubstituted sugar adsorption since if they were involved, as suggested by Lynch *et al.* (1956) among others, then methylation should reduce adsorption. However, he suggests that stronger H-bonds may be formed between O atoms on the clay surface and the C atoms of the methoxyl groups than to other O atoms.

Greenland (1963) proposed a model of organic polymer adsorption to clay in which uncharged molecules were visualized as forming a "coat of paint," with many adsorbed segments, and negatively charged polymers as forming a "string of beads," with few adsorbed segments and many loops into solution. The adsorbed segments may be uncharged with the negatively charged segments being repelled from the surface. The number of adsorbed segments or trains and the distribution and availability of functional groups on organic molecules are important in their interaction with inorganic constituents (Deuel, 1960), and it may be that only a few functional groups are involved in interactions. If a polymer is adsorbed by only a few segments, the strength of the interparticle bond is small (Parfitt and Greenland, 1970).

Martin (1971) summarized the binding activity of polysaccharides as being due to:

1. their length and linear structure, allowing them to bridge spaces between oil particles;

2. their flexibility, allowing many points of contact so that van der Waals forces can be more effective;
3. the number of hydroxyl groups present, so that hydrogen bonds can form; and
4. the number of acid groups present, allowing ionic bonding through di- and trivalent ions.

Greenland (1972) commented that most of the work carried out to elucidate the mechanisms of the adsorption of polymers (both naturally occurring and synthetic) to soil has been done using dispersed systems. In the soil there may be problems in the accessibility of polymers to soil surfaces. Since the effectiveness of polymers in increasing aggregate strength depends on their ability to form interparticle bonds and their ability to penetrate aggregates, their movement through pores may be limiting. For an artificial soil-stabilizing polymer (PVA), the extent of penetration into aggregates was lowest for the highest molecular weight polymer (Carr and Greenland, 1972). Since they observed an increase in aggregate stability at points where no polymer retention could be measured, they concluded that it may be necessary to stabilize only the weakest pores within the system.

The mechanisms of polymer adsorption to clays have been deduced mainly by using pure polymers and pure clays. Soil clay surfaces are very different. The clays are mixtures of different types and may be inter-stratified and have surfaces that are far from clean. Mono- and divalent cations predominate on the cation exchange complex and in solution, whereas many workers have used clays homoionic to polyvalent cations. There is, however, some evidence from experiments using soil to support the various theories. Among others, Guidi *et al.* (1978) found that when polysaccharides of increasing molecular weight, produced in their experiments by *Leuconostoc mesenteroides*, were added to soil aggregates, stability was increased to a greater extent by those with higher molecular weights. By shaking soils with Na-resins, Edwards and Bremner (1967) deduced that polyvalent metals were involved in particle bonding because substituting Na for other cations led to the disruption of aggregates. They postulated that the basic structural units consisted of clay–polyvalent metal–organic matter complexes. Further evidence for the role of poly-valent ions in binding organic polymers in soils comes from the work of Martin and Richards (1963). The binding action of some microbial and plant polysaccharides was reduced if they were allowed to react with Fe and Al ions before being added to soil. Presumably if the active sites on the polymer are already occupied by these ions, they cannot participate in binding through polyvalent ions in the soil.

Martin and Aldrich (1955) found that the binding action of polymers containing uronic acid groups increased in acid conditions as the concentration of the uronic acid groups increased. Binding action decreased with increasing base saturation, especially in the presence of K and Ca ions. This was explained by Fe or Al ions in acid soil either acting as bridges between the polymer and the soil surfaces, so increasing binding in acid soils, or reducing the negative charge on the polymer, so favoring the formation of hydrogen bonds. However, Clapp *et al.* (1962) concluded that the presence of carboxyl groups was not necessary for binding polysaccharides in soil.

Tisdall and Oades (1982) suggest that polysaccharides are transitory binding agents. If their role in aggregate stabilization is more than ephemeral, then either they must be continually replenished with products from bacteria or they must be protected from microbial decomposition in some way. Lasik *et al.* (1978) found that all of the bacterial polysaccharides they looked at could be decomposed by some bacteria. Martin *et al.* (1965) found differences in the aggregating abilities of bacterial polysaccharides and in their decomposition. The polysaccharides they used were not readily decomposed by the majority of the soil population, possibly because some of them inhibit growth and sporing of some bacteria. The polysaccharide of *Chromobacterium violaceum*, found by Corpe (1960) to be very resistant to decomposition, was found by Martin and Richards (1963) to have a greater and more persistent binding action than the other polysaccharides they tested. Using $^{14}$C-labeled bacterial polysaccharides, Martin *et al.* (1974) found that the rate of decomposition of polymers varied greatly.

The resistance of bacterial polymers to decomposition may be a function of their structure or a function of their solubility or ability to form complexes. The presence of metal cations may induce resistance. Metal ions may change the ability of polysaccharides to react with clays. Martin *et al.* (1972) found that the presence of metal ions usually reduced the decomposition of polymers and that they affected the binding abilities of the polymer. Polysaccharides with uronic acid groups or *cis*-hydroxyls at the 2- or 3-C position may be able to form salts or complexes with metals, and complex formation may affect the rate of decomposition of the polymer (Martin *et al.*, 1972). For three bacterial or yeast polysaccharides, they found that complexing with Fe, Al, Cu, or Zn salts reduced their decomposition, with some anomalies, the effect depending on the polymer and the metal.

It is possible that polysacchardies may be protected from degradation by other chemical means. When humic substances were adsorbed to soil, before glucose was added and the soil incubated, persistently stable aggregates were produced (Swift and Cheney, 1979), suggesting that the humic materials were able to confer long-term stability. Griffiths and Burns (1972) found that phenolic substances (tannic acid) prolonged aggregate stabilization by a polysaccharide if the phenolic material interacted with the aggregate after it had been formed, suggesting a physical as well as a chemical function. Phenolic units present in humic acids may originate from fungal decomposition of lignin (Martin and Haider, 1971). Covalent, hydrogen, and ionic bonding are all involved in tanning reactions (Cheshire, 1977).

Rather than being due to their chemistry or structure, the resistance of polysaccharides to decomposition may be due to their unavailability to microorganisms. Substrates may be protected from microbial decomposition by virtue of their position within aggregates. During the formation of aggregates, polysaccharides could become incorporated within aggregates and consequently be physically inaccesible to microorganisms, although not necessarily to enzymes. Cheshire *et al.* (1983) suggested that since there was a direct link between the residual carbohydrate in soil after periodate oxidation and its degree of aggregation, then one could be the cause of the other, i.e., that the carbohydrate that was protected from degradation was in part responsible for aggregate stability. Using transmission electron microscopy to observe thin sections of soils stained to highlight acidic polysaccharides, Foster (1981a,1981b) found threads of fibrous material spanning submicron pores. By using various heavy metal stains, he observed carbohydrate in micropores and between clay platelets, where it may be protected from microbial attack. Using $^{14}$C-labeled glucose and starch added before or after sterile artificial soil aggregates were prepared and inoculated, Adu and Oades (1978a,b) found that starch was protected from attack if it was in the micropores of a fine sandy loam. The time taken for periodate oxidation of a dextran from *Leuconostoc* sp. adsorbed onto a Na-saturated montmorillonite indicated that it was protected in some way from oxidation (Clapp *et al.*, 1968). In soil, the addition of bentonite to artificial soil aggregates inoculated and incubated with *Arthrobacter* sp. increased stability and the length of time it was sustained (Dabek-Szreniawska, 1974). The bentonite may have led to the formation of resistant complexes that protected soil binding components from degradation.

The mechanisms by which organic molecules adsorb to soil particles have become better understood in recent years, but more precise determinations of their structures and solution configurations are needed, especially of the more complex polysaccharides that exist in soil. Microbial polysaccharides, while being effective in aggregation, are often easily decomposed in the soil, although some appear to be protected, either chemically or physically.

## B. Adsorption of Cells to Soil Surfaces

Much work has been done on the adsorption of bacterial cells to surfaces, particularly by Marshall (1971, 1980) in relation to soils. He defined three interactions between microrganisms and soil particles:

1. sorption between microorganisms and surfaces of large soil particles;

2. sorptive interactions between cells and soil particles of like size; and
3. sorption of very small particles to surfaces of microorganisms.

Fehrmann and Weaver (1978) found evidence of adsorption of *Rhizobium meliloti* to silt particles after incubation, but the firm attachment of cells to soil that is found in field soils could be duplicated in the laboratory only after prolonged incubation of autoclaved soils inoculated with natural soil or suspensions of microorganisms (Balkwill and Casida, 1979). Electron and light microscopy have revealed that the distribution of bacteria on soil surfaces is not homogeneous (Gray, 1967).

Electrophoretic mobility measurements have indicated that clay particles can adsorb to cells and vice versa. Clay particles have a higher electrophoretic mobility than bacterial cells, so when they are mixed, changes in the mobility of the components reflect an interaction between cells and particles. Lahav (1962) showed that adsorption of bentonite on *Bacillus subtilis* was reversible. In mixtures of the two components, electrophoretic mobility increased with increasing concentration of clay, indicating that clay adsorbed to the cells. When the concentration of free clay was reduced, mobility returned to that of the bacterial cells. Cooper and Morgan (1979) found that, when mixed with cells of *Escherichia coli*, the mobility of allophane came closer to that of a cell, indicating that cells adsorbed to the clay; adsorption was confirmed by an increase in the number of larger particle sizes as measured by a Coulter Counter (electronic particle counter).

The mechanisms by which bacteria can adsorb to soil surfaces are probably similar to those involved in the interaction of organic polymers to surfaces, since the various functional groups present on organic polymers can be present at bacterial cell surfaces, and extracellular polymers produced by bacteria may provide bridges to the soil surfaces. Saturating cation and pH again have an effect on the interaction. Sorptive interactions between *E. coli* and allophane, as measured by a Coulter Counter, were affected by the saturating cation, more cells adhering per Cu-saturated clay particle than per Na-saturated clay particle, and by pH, with greater adsorption occurring at low pH than in neutral conditions (Cooper and Morgan, 1979). Santoro and Stotzky (1968) found that there was very little adsorption between bacteria and Na-saturated montmorillonite but that there was a shift to larger particle sizes, i.e., adsorption took place, in the presence of polyvalent cations. In the presence of monovalent cations, the greatest adsorption of bacteria was at pH 2 to 4. They suggested that the electrokinetic potential of the particles is reduced in the presence of polyvalent cations, allowing hydrogen bonding and van der Waals forces to be involved in the sorptive reaction. The positive charge on clay edges also increases at low pH, perhaps contributing to the observed adsorption.

One mechanism of adsorption is through functional groups on the surfaces of cells. Using electrophoretic techniques and microscope observation, Marshall (1969a,1969b) investigated the interaction between cells of different rhizobial strains and fine clay that had been treated with $(Na_2PO_3)_6$ to block positive edge charges. There was no increase in the electrophoretic mobility of the cells of one strain, suggesting that clay adsorption to cells that had carboxyl-type surfaces was through positive charges on the clay edges. Limited clay adsorption occurred to cells with more complex amino-carboxyl surfaces, suggesting that positively charged amino groups may interact with negatively charged clay surfaces.

In addition, organelles such as flagellae and pili or fimbrae may play a role in attachment, at least in the initial stages. Nikitin (1964) has observed pili on bacteria isolated from soil, and Zvyagintsev et al. (1969), using a scanning electron microscope to observe morphological features associated with adsorption, suggested that fimbrae were important, although specific attachment structures have rarely been seen by other workers.

If bacteria can interact with soil particles in these ways, what are the consequences for aggregate stabilization? Bacteria surrounded by clay platelets could form a microaggregate (Tisdall and Oades, 1982). When bacteria die they may remain in situ, protected from degradation by the surrounding clay platelets. Scanning electron microscopy of soils by Kilbertus et al. (1977) revealed bacterial colonies in polysaccharide material surrounded by clay particles. In another study, the remains of bacteria and their capsules could be seen as organic matter surrounded by clay (Foster, 1978).

The adhesion of soil particles to fungal hyphae may have important consequences for aggregate stabilization. Fungi tend to produce lower molecular weight polymers than bacteria, their lower molecular weight indicating that they are likely to be less effective soil binding agents than those of bacteria. However, adherence of soil particles to hyphae, indicating the presence of binding agents, has been noted by Aspiras et al. (1971b), Clough and Sutton (1978), and Tisdall and Oades (1979). Using staining techniques, Clough and Sutton (1978) showed that this material contained polysaccharide, although other microorganisms were in association with the hyphae so that the polysaccharide may not necessarily have been produced by the hyphae. Inoculation of aggregates with media in which fungi had been grown showed that stability was a function of substances closely associated with the cells, rather than something that diffused away from the hyphae (Clough and Sutton, 1978).

Fungi may be effective aggregate stabilizers because the spread of hyphae between aggregates and into large pores distributes their associated binding agents throughout the soil. Dead hyphae retain their strength and remain firmly attached to soil particles (Bond and Harris, 1974) so that even after the hyphae die and break up, they may still form the center of small aggregates.

## C. Interactions Between Groups of Microorganisms or with Roots

Interactions between different microbial groups have been noted. McCalla (1946) found that the stability produced by bacteria increased in the presence of other groups. Mixed cultures of fungi or actinomycetes gave better aggregation than soil inoculum (Swaby, 1949), perhaps because the presence of antagonistic bacteria in soil inoculum could decrease the effectiveness of other organisms. Bailey *et al.* (1973) noted that aggregation of loess by algae was enhanced by the presence of their associated bacteria and fungi.

Gel'tser (1940) suggested that the decomposition of fungal hyphae yields products that are better aggregate stabilizers than those from most other sources, and Low and Stuart (1974) suggested that although neither roots nor fungal hyphae persist for long in soil, they may act as scaffolding from which the products of bacterial action could form transient binding agents. Several workers have observed bacteria in the amorphous material around fungal hyphae (Clough and Sutton, 1978; Tisdall and Oades, 1979). Using low temperature scanning electron microscopy, Campbell and Porter (1982) revealed more clearly the relationship between fungal hyphae and bacterial cells, with bacteria being present in the water film around hyphae.

However, although the incubation of dead sterile hyphae in soil increased aggregate stability and supported vigorous growth of filamentous micro-organisms, aggregates were rarely more persistent then those produced by the incubation of glucose (Griffiths and Jones, 1965). Martin *et al.* (1959) found that fungal material decomposed more quickly in soil than many plant residues, although hyphae containing more lignin took longer to decompose than others. During the decomposition of lignin-containing fungal hyphae, any bacterial polysaccharides could be protected against further degradation by phenolic units produced as by-products of decomposition (Martin and Haider, 1971). McHenry and Russell (1944) attributed increases in aggregate stability to the decomposition of microbial wastes and dead cells. They found that there were two maxima in aggregate stability and suggested that the first came from the decomposition of readily available organic materials and the second from the decomposition of cells and wastes. Their results are substantiated by those of Adu and Oades (1978b). When $^{14}C$-labeled starch and glucose were added to soil there were two peaks in $^{14}CO_2$ production if both bacteria and fungi were present in the soil. They suggested that the first was due to the activity of fungi and the second to bacteria metabolizing dead fungal materials and waste products.

Plant roots have been shown to increase the stability of surrounding aggregates (Tisdall and Oades, 1979; Reid and Goss, 1980). Sutton and Sheppard (1976) found that hyphae of *Glomus* sp. (an endomycorrhizal fungus of beans) were a major factor in the aggregation of a Canadian dune soil by beans (*Phaseolus vulgaris*) and that aggregation was enhanced in the presence of other organisms, possibly through stimulation of fungal growth. Tisdall and Oades (1979) concluded that the efficiency of ryegrass (*Lolium perenne*) in aggregate stabilization was due to the large vesicular–arbuscular mycorrhizal fungal population it supported, because hyphal length was related to aggregate stability, and they later suggested that the organic materials that stabilize aggregates are decomposing roots and fungal hyphae (Tisdall and Oades, 1980). Roots themselves release large amounts of organic materials into soil, seen as a slime layer around the root (Oades, 1978; Campbell and Porter, 1982), together with sloughed-off root cells and root hairs. These can be decomposed by bacteria, the numbers of which are increased in the rhizosphere. Webley et al. (1965) found that the root surfaces of three grasses had a higher proportion of bacteria that were able to produce capsules and/or slime material, compared with either rhizosphere or non-rhizosphere soil.

Therefore, several interacting mechanisms may be responsible for aggregate stabilization around roots. The roots themselves can move particles together, and their localized drying of the soil could stabilize aggregates to some extent (Allison, 1968). The presence of fungi, possibly vesicular–arbuscular mycorrhiza, could mechanically bind soil particles together, with stabilization being enhanced by polymers produced either directly by the fungus or by bacteria associated with the hyphae. Bacteria at the root surface would be in an ideal position to utilize root residues to produce effective soil binding agents.

# V. Applications

Now that the role of microorganisms in soil aggregation processes is becoming clearer, the question can be posed as to whether there is any prospect for manipulating the soil microbial population to promote aggregate stability. The following serve to illustrate a few means by which this might be achieved.

## A. The Rhizosphere Effect

Grassland soils generally have very stable soil structures (Low, 1955; Robinson and Jaques, 1958; Clarke et al., 1967; Reid and Goss, 1980). This is because there is a much greater root biomass and hence associated rhizosphere microbial biomass produced under grass (Lynch, 1981b). The root produces mucilage, and the associated microorganisms modify the products released by roots to produce their own mucilage; together the mucilages form mucigel. These materials are both largely polysaccharide in nature and hence have the capacity to stabilize soil aggregates. Both forms of mucilage are probably important in stabilizing soils, but the relative importance of each is unclear. The scanning electron microscope demonstrates bacterial cells embedded in the mucigel; this frequently takes place by end-on "docking." The transmission electron microscope has demonstrated clay particles adhered to the polysaccharide surrounding rhizo-sphere bacterial (Faull and Campbell, 1979).

Plants vary in the amounts and types of organic materials released by their roots; correspondingly the rhizosphere population varies between plant species. The inoculation of seeds or seedling roots with polysac-charide-producing microorganisms is unlikely to offer a realistic economic prospect if their only function is to improve soil structure. However, a more realistic approach is to consider growing break crops that have naturally effective soil stabilizing rhizospheres. At Saxmundham (Suffolk, England) the sandy soil is particularly poorly structured, and crop yields are extremely low (Cooke and Williams, 1972). However, following a crop of lucerne (*Medicago sativa*), the structure is improved and subsequent arable crop yields are increased. In pot experiments, Reid and Goss (1981) also found that the growth of lucerne increased aggregate stability. Is there something especially significant about the lucerne rhizosphere? Reid and Goss (1980) demonstrated that the growth of ryegrass roots improved soil aggregate stability in a silt loam soil but, surprisingly, the growth of maize (*Zea mays*) decreased aggregate stability. We have been unable to repeat the results with maize, perhaps owing to differences in experimental procedure. It would be unwise to generalize on the effects at this stage, clearly pointing to the need for more research.

## B. Crop Residues

Straw and other crop residues can provide substrates for microorganisms to produce agents to stabilize soil aggregates (Gilmour *et al.*, 1948). Although it has been indicated that the inoculation of straw with microorganisms does not improve its aggregate stabilizing effect on soil and volcanic ash over that induced by the natural microflora present (Lynch and Elliott, 1983), recent work (S.J. Chapman and J.M. Lynch, un-published observations) has shown more promise.

The aggregate stabilizing effect from straw degradation decreased when straw N content increased in the range 0.25 to 1.09% w/w (Elliott and Lynch, 1984). Hot water extracts of the polysaccharide from decomposed straw was composed mainly of galactose, glucose, and mannose, with smaller quantities of arabinose, xylose, rhamnose, fucose, and ribose, indicating a mainly microbial origin (Chapman and Lynch, 1984). The polysaccharide produced a dose-response curve for the stabilization of volcanic ash.

In the U.S., straw residues are used in conservation tillage systems, contributing to an improvement in soil structure and reducing erosion. In Britain, straw is generally burned and, although soil structures are usually relatively stable, clearly caution should be exercised where there are inherent soil structural problems.

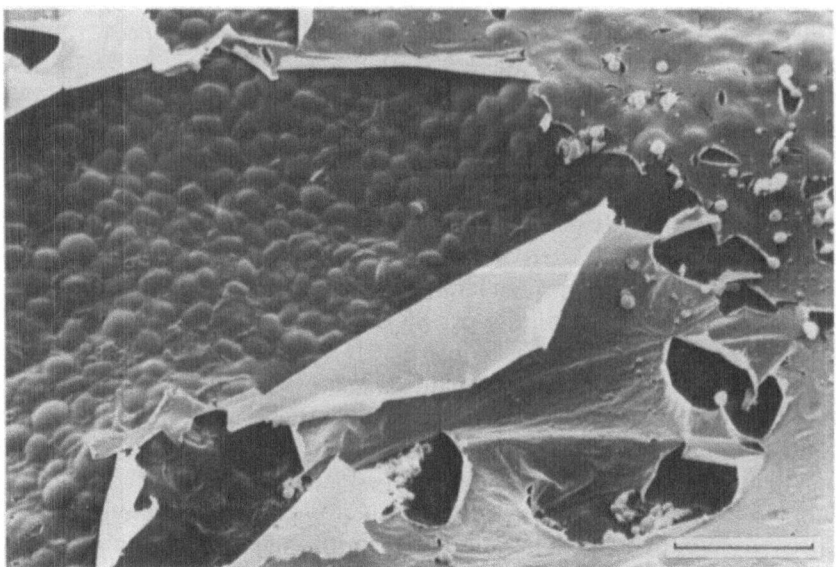

**Figure 5.** Algal crust on the surface of undisturbed soil. A few diatoms are also evident among the predominantly unicellular algae (bar marker 30μm). Photograph courtesy of Dr. J. Sargent, AFRC Weed Research Organisation, Oxford.

## C. Algae as Soil Conditioners

Algae are frequently apparent in the surface crusting of dry soils following rainfall (Figure 5). There thus seems potential to inoculate soil with algae to increase this effect and stabilize such soils so that they might retain more water. Metting and Rayburn (1983) inoculated a loamy fine sand and a silt loam with a mass-cultured *Chlamydomonas mexicana* and improved soil aggregate stability. Moisture was the most likely factor controlling growth and polysaccharide production by the algal inoculum, and no benefit could be found in a dryland silt loam. Clearly the usefulness of this approach is likely to be restricted to irrigated soils where there is an abundant source of light, but, within those limits, there appears some commercial prospect.

## D. Animal and Domestic Wastes

It is generally considered that the addition of organic matter in the form of animal wastes to soil will improve its structure. Relatively few studies have been aimed at quantifying this, and the contribution of the microbial components of the wastes does not seem to have been assessed. Some researchers claim that composted domestic refuse improves structure when added to soil, but quantification of this appears even more scanty. In the latter situation a potential environmental problem can occur from the simultaneous introduction of heavy metals.

# VI. Conclusions

The study of microbial effects on soil aggregate stability and the manipulation of these activities is a target of soil biotechnology (Lynch, 1983). Modification of existing cropping sequences and practices might be considered when more laboratory and field evaluation has been carried out. The prospect of direct manipulation of microbial activities by inoculation is exciting but might only become economic when the inoculum is multifunctional, for example, also providing plant nutrients or acting as a biocontrol agent against disease.

# References

Acton, C.J., D.A. Rennie, and E.A. Paul. 1963. The relationship of polysaccharides to soil aggregation. *Can. J. Soil Sci.* 43:201–209.

Adu, J.K., and J.M. Oades. 1978a. Physical factors influencing decomposition of organic materials in soil aggregates. *Soil Biol. Biochem.* 10:109–115.

Adu, J.K., and J.M. Oades. 1978b. Utilization of organic materials in soil aggregates by bacteria and fungi. *Soil Biol. Biochem.* 10:117–122.

Aldrich, D.G. 1948. The effect of soil acidification on some physical and chemical properties of three irrigated soils. *Soil Sci. Soc. Am. Proc.* 13:191–196.

Allison, F.E. 1968. Soil aggregation—some facts and fallacies as seen by a microbiologist. *Soil Sci.* 106:136–144.

Aspiras, R.B., O.N. Allen, G. Chesters, and R.F. Harris. 1971a. Chemical and physical stability of microbially stabilised aggregates. *Soil Sci. Soc. Am. Proc.* 35:283–286.

Aspiras, R.B., O.N. Allen, R.F. Harris, and G. Chesters. 1971b. Aggregate stabilisation by filamentous micro-organisms. *Soil Sci.* 112:282–284.

Bailey, D., A.P. Mazurak, and J.R. Rosowski. 1973. Aggregation of soil particles by algae. *J. Phycol.* 9:99–101.

Balkwill, D.L., and L.E. Casida. 1979. Attachment to autoclaved soil of bacterial cells from pure cultures of soil isolates. *App. Environ. Microbiol.* 37:1031–1037.

Batey, T. 1974. Soil structure: its effect on crop yield. In: *Forage on the arable farm.* pp. 5–11. Occasional Symposium No. 7, British Grassland Society.

Bobbit, J.M. 1956. Periodate oxidation of carbohydrates. *Adv. Carb. Chem.* 11:1–43.

Bond, R.D. 1959. Occurrence of microbial filaments in soils. *Nature* 184:744–745.

Bond, R.D., and J.R. Harris. 1964. The influence of the microflora on physical properties of soils. I. Effects associated with filamentous algae and fungi. *Austr. J. Soil Res.* 2:111–122.

Brown, L.R. 1981. World population growth, soil erosion and food security. *Science* 214:995–1002.

Burns, R.G. 1977. *The soil microenvironment: aggregates, enzymes and pesticides.* CNR, Lab. Chimica Terrene Conference 5(1).

Burns, R.G. 1979. Interactions of micro-organisms, their substrates and their products with soil surfaces. In: D.C. Ellwood, J. Melling, and P. Rutter (eds.), *Adhesion of micro-organisms to surfaces.* pp. 109–138. Academic Press, London.

Burns, R.G. 1983. Extracellular enzyme-substrate interactions in soil. In: J.H. Slater, R. Whittenbury, and J.W.T. Wimpenny (eds.), *Microbes in their natural environments.* pp. 249–298. Cambridge University Press, Cambridge.

Campbell, R. 1983. *Microbial ecology.* 2d ed. Blackwell Scientific Publications, Oxford.

Campbell, R., and R. Porter. 1982. Low temperature scanning electron microscopy of micro-organisms in soil. *Soil Biol. Biochem.* 14:241–245.

Carr, C.E., and D.J. Greenland. 1972. Preliminary result of an empirical study of the movement of polymers through soil and their effect on dispersion of clay from aggregates. In: M. De Boodt (ed.), *Proc. symposium on fundamentals of soil conditioning.* pp. 982–992. State University of Ghent, Belgium, Faculty of Agricultural Sciences.

Chapman, S.J., and J.M. Lynch. 1984. A note on the formation of microbial polysaccharide from wheat straw decomposed in the absence of soil. *J. Appl. Bacter.* 56:337–342.

Cheshire, M.V. 1977. Origins and stability of soil polysaccharide. *J. Soil Sci.* 28:1–10.

Cheshire, M.V. 1979. *Nature and origin of carbohydrates in soil.* Academic Press, London.

Cheshire, M.V., and C.M. Mundie. 1981. The distribution of labelled sugars in soil particle size fractions as a means of distinguishing plant and microbial carbohydrate residues. *J. Soil Sci.* 32:605–618.

Cheshire, M.V., C.M. Mundie, and H. Shepherd. 1969. Transformation of $^{14}$C-glucose and starch in soil. *Soil Biol. Biochem.* 1:117–130.

Cheshire, M.V., G.P. Sparling, and C.M. Mundie. 1983. Effect of periodate treatment of soil on carbohydrate constituents and soil aggregation. *J. Soil Sci.* 34:105–112.

Chesters, G., O.J. Attoe, and O.N. Allen. 1957. Soil aggregation in relation to various soil constituents. *Soil Sci. Soc. Am. Proc.* 21:272–277.

Clapp, C.E., and W.W. Emerson. 1965. The effect of periodate oxidation on the strength of soil crumbs. I. Qualitative studies. *Soil Sci. Soc. Am. Proc.* 29:127–130.

Clapp, C.E., R.J. Davis, and S.H. Waugaman. 1962. The effect of rhizobial polysaccharides on aggregate stability. *Soil Sci. Soc. Am. Proc.* 26:466–469.

Clapp, C.E., A.E. Olness, and D.J. Hoffmann. 1968. Adsorption studies of a dextran on montmorillonite. *Trans. 9th Int. Congr. Soil Sci.* (Adelaide) 1:627–634.

Clarke, A.L., D.J. Greenland, and J.P. Quirk. 1967. Changes in some physical properties of the surface of an impoverished red-brown earth under pasture. *Austr. J. Soil Res.* 5:59–68.

Clough, K.S., and J.C. Sutton. 1978. Direct observation of fungal aggregates in sand dune soil. *Can. J. Microbiol.* 24:333–335.

Cooke, G.W., and R.J.B. Williams. 1972. Problems with cultivation and soil structure at Saxmundham. *Rothamsted Report*, 1971, Part 2. pp. 122–142.

Cooper, A.B., and H.W. Morgan. 1979. Interactions between *Escherichia coli* and allophane. I. Adsorption. *Soil Biol. Biochem.* 11:221–226.

Corpe, W.A. 1960. The extra-cellular polysaccharide of gelatinous strains of *Chromobacter violaceum. Can. J. Microbiol.* 6:153–163.

Dabek-Szreniawska, M. 1972. The influence of carbon on the production of slime material by some azotobacter strains. *Polish J. Soil Sci.* 5:59–67.

Dabek-Szreniawska, M. 1974. The influence of *Arthrobacter* sp. on the water stability of soil aggregates. *Polish J. Soil Sci.* 7:169–179.

Deuel, H. 1960. Interactions between organic and inorganic soil constituents. *Trans. 7th Intl. Congr. Soil Sci.* (Madison, WI). pp. 38–52.

Dormaar, J.F., and V.J. Pittman. 1980. Decomposition of organic residues as affected by various dryland spring wheat–fallow rotations. *Can. J. Soil Sci.* 60:97–106.

Douglas, J.T., and M.J. Goss. 1982. Stability and organic matter content of surface soil aggregates under different methods of cultivation and in grassland. *Soil Till. Res.* 2:155–175.

Dudman, W.F. 1977. The role of surface polysaccharides in natural environments. In: I.W. Sutherland (ed.), *Surface carbohydrates of the prokaryotic cell.* pp. 357–414. Academic Press, London.

Edwards, A.P., and J.M. Bremner. 1967. Microaggregates in soil. *J. Soil Sci.* 18:64–73.

Elliott, L.F., and J.M. Lynch. 1984. The effect of available carbon and nitrogen in straw on soil and ash aggregation and acetic acid production. *Plant and Soil.* 78:335–343.

Emerson, W.W. 1954. The determination of the stability of soil crumbs. *J. Soil Sci.* 5:233–250.

Emerson, W.W. 1959. The structure of soil crumbs. *J. Soil Sci.* 10:235–244.

Faull, J.L., and R. Campbell. 1979. Ultrastructure of the interaction between the take-all fungus and antagonistic bacteria. *Can. J. Bot.* 57:1800–1808.

Fehrmann, R.C., and R.W. Weaver. 1978. Scanning electron microscopy of *Rhizobium* sp. adhering to fine silt particles. *Soil Sci. Soc. Am. J.* 42:279–281.

Finch, P., M.H.B. Hayes, and M. Stacey. 1971. The biochemistry of soil polysaccharides. In: A.D. McLaren and J. Skujins (eds.), *Soil biochemistry*. vol. 2. pp. 257–319. Marcel Dekker, New York.

Fletcher, M.M., M.J. Latham, J.M. Lynch, and P.R. Rutter. 1980. The characteristics of interfaces and their role in microbial attachment. In: R.C.W. Berkeley, J.M. Lynch, J. Melling, P.R. Rutter, and B. Vincent (eds.), *Microbial adhesion to surfaces*. pp. 67–78. Ellis Horwood, Chichester.

Forsyth, W.G.C., and D.M. Webley. 1949. The synthesis of polysaccharides by bacteria isolated from soil. *J. Gen. Microbiol.* 3:395–399.

Foster, R.C. 1978. Ultramicroscopy of some South Australian soils. In: W.W. Emerson, R.C. Bond, and A.R. Dexter (eds.), *Modifications of soil structure*. pp. 103–109. John Wiley and Sons, London.

Foster, R.C. 1981a. Polysaccharides in soil fabrics. *Science* 214:665–667.

Foster, R.C. 1981b. Localization of organic materials *in situ* in ultrathin sections of natural soil fabrics using cytochemical techniques. *International Working Group on Submicroscopy of Undisturbed Soil Materials*. pp. 309–317. Waglningen, Pudoc Press.

Gati, F. 1982. Use of organic materials as soil amendments. In: *Organic materials and soil productivity in the Near East*. FAO Soils Bull. No. 45. pp. 87–105. FAO, Rome.

Gaur, A.C., and R.V.S. Rao. 1975. Note on the isolation of bacterial gums and their influence on soil aggregate stabilisation. *Ind. J. Agr. Sci.* 45:86–89.

Gel'tser, F.Y. 1940. The significance of micro-organisms in the formation of humus (Summary). *Soils and Fert.* (1943)7:119–121.

Geoghegan, M.J. 1950. Aggregate formation in soil. Influence of some microbial metabolic products and other substances on aggregation of soil particles. *Trans. Intl. Congr. Soil Sci.* (Amsterdam) 1:198–201.

Geoghegan, M.J., and E.R. Armitage. 1949. Influence of some lipoidal substances on aggregate formation in soils. *Nature* 163:29–30.

Geoghegan, M.J., and R.C. Brian. 1946. Influence of bacterial polysaccharides on aggregate formation in soils. *Nature* 158:837–838.

Geoghegan, M.J., and R.C. Brian. 1948. Aggregate formation in soil. I. Influence of some bacterial polysaccharides on the binding of soil particles. *Biochem. J.* 43:5–13.

Gilmour, C.M., O.N. Allen, and E. Truog. 1948. Soil aggregation as influenced by the growth of mold species, kind of soil, and organic matter. *Soil Sci. Soc. Am. Proc.* 13:292–296.

Giovannini, G., and P. Sequi. 1976. Iron and aluminum as cementing substances of soil aggregates. II. Changes in stability of soil aggregates following extraction of iron and aluminum by acetylacetone in a non-polar solvent. *J. Soil Sci.* 27:148–153.

Giovannini, G., S. Lucchesi, and S. Cervelli. 1983. Water-repellent substances and aggregate stability in hydrophobic soil. *Soil Sci.* 135:110–113.

Gray, T.R.G. 1967. Stereoscan electron microscopy of soil micro-organisms. *Science* 155:1668–1670.

Greenland, D.J. 1956a. The adsorption of sugars by montmorillonite. I. X-ray studies. *J. Soil Sci.* 7:319–328.

Greenland, D.J. 1956b. The adsorption of sugars by montmorillonite. II. Chemical studies. *J. Soil Sci.* 7:329–334.

Greenland, D.J. 1963. Adsorption of polyvinyl alcohols by montmorillonite. *J. Colloid Sci.* 18:647–664.

Greenland, D.J. 1965a. Interaction between clays and organic compounds in soils. I. Mechanisms of interaction between clays and defined organic compounds. *Soils Fert.* 28:415–426.

Greenland, D.J. 1965b. Interaction between clays and organic compounds in soils. II. Adsorption of soil organic compounds and its effect on soil properties. *Soils Fert.* 28:521–532.

Greenland, D.J. 1970. Sorption of organic compounds by clays. In: *Sorption and transport processes in soil.* Monograph No. 37. pp. 79–91. Society of Chemical Industry, London.

Greenland, D.J. 1972. Interactions between organic polymers and inorganic soil particles. In: M. De Boodt (ed.), *Proc. symposium on fundamentals of soil conditioning.* pp. 897–914. State University of Ghent, Belgium, Faculty of Agricultural Sciences.

Greenland, D.J., G.R. Lindstrom, and J.P. Quirk. 1961. Role of polysaccharides in stabilisation of natural soil aggregates. *Nature* 191:1283–1284.

Greenland, D.J., G.R. Lindstrom, and J.P. Quirk. 1962. Organic materials which stabilise natural soil aggregates. *Soil Sci. Soc. Am. Proc.* 26:366–371.

Griffiths, E., and R.G. Burns. 1972. Interaction between phenolic substances and microbial polysaccharides in soil aggregation. *Plant and Soil* 36:599–612.

Griffiths, E., and D. Jones. 1965. Microbiological aspects of soil structure. I. Relationships between organic amendments, microbial colonization and changes in aggregate stability. *Plant and Soil* 23:17–33.

Guidi, G., M. Pagliai, G. Petruzzelli, and R. Aringhieri. 1978. Changes in some physical properties of clay soils induced by dextrans. *Z. Pflanzen. Bodenk.* 141:367–377.

Hamblin, A.P. 1980. Changes in aggregate stability and associated organic matter properties after direct drilling and ploughing on some Australian soils. *Austr. J. Soil Res.* 18:27–36.

Hansen, L. 1982. Problems of soil structure in monoculture of spring barley. In: D. Boels, D.B. Davies, and A.E. Johnston (eds.), *Soil degradation* pp. 87–94. A.A. Balkema, Rotterdam.

Harris, R.F., G. Chesters, and O.N. Allen. 1966a. Dynamics of soil aggregation. *Adv. Agron.* 18:107–169.

Harris, R.F., G. Chesters, and O.N. Allen. 1966b. Soil aggregate stabilisation by the indigenous microflora as affected by temperature. *Soil Sci. Soc. Am. Proc.* 30:205–210.

Harris, R.F., O.N. Allen, G. Chesters, and O.J. Attoe. 1963. Evaluation of microbial activity in soil aggregate stabilisation and degradation by the use of artificial aggregates. *Soil Sci. Soc. Am. Proc.* 27:542–545.

Harris, R.F., G. Chesters, O.N. Allen, and O.J. Attoe. 1964. Mechanisms involved in soil aggregate stabilisation by fungi and bacteria. *Soil Sci. Soc. Am. Proc.* 28:529–532.

Hayes, M.H.B. 1980. Role of natural and synthetic polymers in stabilising soil aggregates. In: R.C.W. Berkeley, J.M. Lynch, J. Melling, P.R. Rutter, and B. Vincent (eds.), *Microbial adhesion to surfaces.* pp. 262–296. Ellis Horwood, Chichester.

Johnston, A.E. 1982. The effects of farming systems on the amount of soil organic matter and its effect on yield at Rothamsted and Woburn. In: D. Boels, D.B. Davies, and A.E. Johnston (eds.), *Soil degradation.* pp. 187–202. A.A. Balkema, Rotterdam.

Keefer, R.F., and J.F. Mortenson. 1963. Biosynthesis of soil polysaccharides. I. Glucose and alfalfa tissue substrates. *Soil Sci. Soc. Am. Proc.* 27:156–160.

Kilbertus, G. 1980. Etude des microhabitats contenus dans les agregats du sol: leur relation avec la biomass bacterienne et la taille des procaryotes presents. *Rev. Ecol. Biol. Sol* 17:543–558.

Kilbertus, G., J. Proth, and F. Magenot. 1977. On the distribution and survival of soil micro-organisms: electron microscope study. *Bull. Acad. Soc. Lorraines Sci.* (Nancy) 16:93–104. Abstract in *Microbiol. Abs.* 14B (1979) No. 3449.

Lahav, N. 1962. Adsorption of sodium bentonite particles on *Bacillus subtilis. Plant and Soil* 17:191–208.

Lasik, Y.A., S.A. Gordiyenko, and L. Kalakhova. 1978. Decomposition of bacterial polysaccharides in soil. *Soviet Soil Sci.* 10:151–153.

Low, A.J. 1955. Improvements in the structural state of soils under leys. *J. Soil Sci.* 6:179–199.

Low, A.J., and P.R. Stuart. 1974. Microstructural differences between arable and old grassland soils as shown in the scanning electron microscope. *J. Soil Sci.* 25:135–143.

Lutz, J.F. 1936. The relation of free iron oxide in the soil to aggregation. *Soil Sci. Soc. Am. Proc.* 1:43–45.

Lynch, D.L., L.M. Wright, and L.J. Cotnoir. 1956. The adsorption of carbohydrates and related compounds on clay minerals. *Soil Sci. Soc. Am. Proc.* 20:6–9.

Lynch, J.M. 1981a. Promotion and inhibition of soil aggregate stabilisation by micro-organisms. *J. Gen. Microbiol.* 126:371–375.

Lynch, J.M. 1981b. Interactions between bacteria and plants in the root environment. In: M.E. Rhodes-Roberts, and F.A. Skinner (eds.), *Bacteria and plants.* pp. 1–23. Academic Press, London.

Lynch, J.M. 1983. *Soil biotechnology: microbiological factors in crop productivity.* Blackwell Scientific Publications, Oxford.

Lynch, J.M., and L.F. Elliott. 1983. Aggregate stabilisation of volcanic ash and soil during microbial degradation of straw. *Appl. Environ. Microbiol.* 45:1398–1401.

Lynch, J.M., and S.J. Pryn. 1977. Interaction between a soil fungus and barley seed. *J. Gen. Microbiol.* 103:193–196.

Marshall, K.C. 1969a. Studies by microelectrophoretic and microscopic techniques of the sorption of illite and montmorillonite to rhizobia. *J. Gen. Microbiol.* 56:301–306.

Marshall, K.C. 1969b. Orientation of clay particles sorbed on bacteria possessing different ionogenic surfaces. *Biochim. Biophys. Acta* 193:472–474.

Marshall, K.C. 1971. Sorptive interactions between soil particles and micro-organisms. In: A.D. McLaren and J. Skujins (eds.), *Soil biochemistry.* Vol. 2. pp. 409–445. Marcel Dekker, New York.

Marshall, K.C. 1980. Adsorption of micro-organisms to soils and sediments. In: G. Bitton and K.C. Marshall (eds.), *Adsorption of micro-organisms to surfaces.* pp. 317–329. John Wiley and Sons, New York.

Martin, J.K. 1975. $^{14}$C labelled material leached from the rhizosphere of plants supplied continuously with $^{14}CO_2$. *Soil Biol. Biochem.* 7:395–399.

Martin, J.P. 1942. The effect of composts and compost materials upon the aggregation of the silt and clay particles of Collington sandy loam. *Soil Sci. Soc. Am. Proc.* 7:218–222.

Martin, J.P. 1945. Micro-organisms and soil aggregation. I. Origin and nature of some of the aggregating substances. *Soil Sci.* 59:163–174.

Martin, J.P. 1946. Micro-organisms and soil aggregation. II. Influence of bacterial polysaccharides on soil structure. *Soil Sci.* 61:157–166.

Martin, J.P. 1971. Decomposition and binding action of polysaccharides in soil. *Soil Biol. Biochem.* 3:33–41.

Martin, J.P., and D.G. Aldrich. 1955. Influence of soil exchangeable cation ratios on the aggregating effects of natural and synthetic soil conditioners. *Soil Sci. Soc. Am. Proc.* 19:50–54.

Martin, J.P., and B.A. Craggs. 1946. Influence of temperature and moisture on the soil aggregating effect of organic residues. *J. Am. Soc. Agron.* 38:332–339.

Martin, J.P., and K. Haider. 1971. Microbial activity in relation to soil humus formation. *Soil Sci.* 111:54–63.

Martin, J.P., and S.J. Richards. 1963. Decomposition and binding action of a polysaccharide from *Chromobacterium violaceum* in soil. *J. Bacteriol.* 85:1288–1294.

Martin, J.P., and S.A. Waksman. 1940. Influence of micro-organisms on soil aggregation and erosion. I. *Soil Sci.* 50:29–47.

Martin, J.P., and S.A. Waksman. 1941. Influence of micro-organisms on soil aggregation and erosion. II. *Soil Sci.* 52:381–394.

Martin, J.P., J.O. Ervin, and S.J. Richards. 1972. Decomposition and binding action in soil of some mannose-containing microbial polysaccharides and their Fe, Al, Zn and Cn complexes. *Soil Sci.* 113:322–327.

Martin, J.P., J.O. Ervin, and R.A. Shepherd. 1959. Decomposition and aggregating effect of fungus cell material in soil. *Soil Sci. Soc. Am. Proc.* 23:217–220.

Martin, J.P., J.O. Ervin, and R.A. Shepherd. 1965. Decomposition and binding action of polysaccharides from *Azotobacter indicus* (*Beijerinckia*) and other bacteria in soil. *Soil Sci. Soc. Am. Proc.* 29:397–400.

Martin, J.P., K. Haider, W.J. Farmer, and E. Fustec-Mathon. 1974. Decomposition and distribution of residual activity of some [14]C-microbial polysaccharides and cells, glucose, cellulose and wheat straw in soil. *Soil Biol. Biochem.* 6:221–230.

Martin, J.P., W.P. Martin, J.B. Page, W.A. Raney, and J.G. De Ment. 1955. Soil aggregation. *Adv. Agron.* 7:1–37.

McCalla, T.M. 1945. Influence of micro-organisms and some organic substances on soil structure. *Soil Sci.* 59:287–297.

McCalla, T.M. 1946. Influence of some microbial groups on stabilizing soil structure against falling water drops. *Soil Sci. Soc. Am. Proc.* 11:260–263.

McCalla, T.M., F.A. Haskins, and E.F. Frolik. 1957. Influence of various factors on aggregation of Peorian loess by micro-organisms. *Soil Sci.* 84:155–161.

McHenry, J.R., and M.B. Russell. 1944. Microbial activity and aggregation of mixtures of bentonite and sand. *Soil Sci.* 57:351–357.

Mehta, N.C., H. Streuli, M. Muller, and H. Deuel. 1960. Role of polysaccharides in soil aggregation. *J. Sci. Food Agr.* 11:40–47.

Metting, B., and W.R. Rayburn. 1983. The influence of a microalgal conditioner on selected Washington soils: an empirical study. *Soil Sci. Soc. Am. J.* 47:682–685.

Meyers, H.E., and T.M. McCalla. 1941. Changes in soil aggregation in relation to bacterial numbers, hydrogen ion concentration and length of time soil was kept moist. *Soil Sci.* 51:189–200.

Newman, R.H., K.R. Tate, P.F. Barron, and M.A. Wilson. 1980. Towards a direct, non-destructive method of characterising soil humic substances using $^{13}$C nuclear magnetic resonance. *J. Soil Sci.* 31:623–631.

Nikitin, D.I. 1964. Use of electron microscopy in the study of soil suspensions and cultures of micro-organisms. *Soviet Soil Sci.* 7:636–641.

Oades, J.M. 1978. Mucilages at the root surface. *J. Soil Sci.* 29:1–16.

Olness, A.E., and C.E. Clapp. 1975. Influence of polysaccharide structure on dextran adsorption by montmorillionite. *Soil Biol. Biochem.* 7:113–118.

Olness, A., and C.E. Clapp. 1976. Influence of polymer structure on stability of dextrans complexed with montmorillonite. In: M. DeBoodt and D. Gabriels (eds.), *Third Intl. symposium on soil conditioning.* pp. 241–252. State University of Ghent, Belgium, Faculty of Agricultural Sciences.

Page, E.R. 1983. Fresh interest in soil conditioners. *Span* 26:10–11.

Parfitt, R.L., and D.J. Greenland. 1970. Adsorption of polysaccharides by montmorillonite. *Soil Sci. Soc. Am. Proc.* 34:862–866.

Peele, T.C. 1940. Microbial activity in relation to soil aggregation. *J. Am. Soc. Agron.* 32:204–212.

Rathore, T.R., B.P. Ghildyal, and R.S. Sachan. 1982. Germination and emergence of soybean under crusted soil conditions. II. Seed environment and varietal differences. *Plant and Soil* 65:73–77.

Reid, J.B., and M.J. Goss. 1980. Changes in the aggregate stability of a sandy loam soil effected by growing roots of a perennial ryegrass (*Lolium perenne*). *J. Sci. Food Agr.* 31:325–328.

Reid, J.M., and M.J. Goss. 1981. Effect of living roots of different plant species on the aggregate stability of two arable soils. *J. Soil Sci.* 32:522–542.

Rennie, D.A., E. Truog, and O.N. Allen. 1954. Soil aggregation as influenced by microbial gums, level of fertility and kind of crop. *Soil Sci. Soc. Am. Proc.* 18:399–403.

Robinson, D.O., and J.B. Page. 1950. Soil aggregate stability. *Soil Sci. Soc. Am. Proc.* 15:25–29.

Robinson, G.S., and W. A. Jacques. 1958. Root development in some common New Zealand pasture plants. X. Effect of pure sowings of some grasses and clovers on the structure of a Tokomaru silt loam. *N. Z. J. Agr. Res.* 1:999–216.

Santoro, T., and G. Stotzky. 1967. Influence of cations on flocculation of clay minerals by microbial metabolites as determined by the electrical sensing zone particle analyzer. *Soil Sci. Soc. Am. Proc.* 31:761–765.

Santoro, T., and G. Stotzky. 1968. Sorption between micro-organisms and clay minerals as determined by the electrical sensing zone particle analyzer. *Can. J. Microbiol.* 14:299–307.

Sequi, P. 1978. Soil structure—an outlook. *Agrochimica* 22:403–425.

Skinner, F.A. 1979. Rothamsted studies of soil structure. VII. The effects of incubation on soil aggregate stability. *J. Soil Sci.* 30:473–481.

Sparling, G.P., M.V. Cheshire, C.M. Mundie, and S. Murayama. 1981. The transformation of [14]C-labelled glucose in sterilised soil inoculated with selected micro-organisms. *Rev. Ecol. Biol. Sol* 18:447–457.

Stotzky, G., and V. Bystricky. 1969. Electron microscopic observations of surface interactions between clay minerals and micro-organisms. *Bacteriol. Proc.* A93.

Sutherland, I.W. 1977. Bacterial exopolysaccharides—their nature and production. In: I.W. Sutherland (ed.), *Surface carbohydrates of the prokaryotic cell.* pp. 27–96. Academic Press, London.

Sutton, J.C., and Sheppard, B.R. 1976. Aggregation of sand-dune soil by endomycorrhizal fungi. *Can. J. Bot.* 54:326–333.

Swaby, R.J. 1949. The relationship between micro-organisms and soil aggregation. *J. Gen. Microbiol.* 3:236–254.

Swift, R.S., and K. Cheney. 1979. The role of soil organic colloids in the formation and stabilisation of soil aggregates. *J. Sci. Food Agr.* 30:329–330.

Swincer, G.D., J.M. Oades, and D.J. Greenland. 1969. The extraction, characterisation and significance of soil polysaccharide. *Adv. Agron.* 21:195–235.

Theng, B.K.G. 1982. Clay—polymer interactions: summary and perspectives. *Clays Clay Min.* 30:1–10.

Tisdall, J.M., and J.M. Oades. 1979. Stabilisation of soil aggregates by the root systems of ryegrass. *Austr. J. Soil Res.* 17:429–441.

Tisdall, J.M., and J.M. Oades. 1980. The effect of crop rotation on aggregation in a red-brown earth. *Austr. J. Soil Res.* 18:423–434.

Tisdall, J.M., and J.M. Oades. 1982. Organic matter and water stable aggregates in soils. *J. Soil Sci.* 33:141–164.

Watson, J.H., and B.J. Stojanovic. 1965. Synthesis and bonding of soil aggregates as affected by microflora and its metabolic products. *Soil Sci.* 100:57–62.

Webley, D.M., R.B. Duff, J.S.D. Bacon, and V.C. Farmer. 1965. A study of the polysaccharide producing organisms occurring in the root region of certain pasture grasses. *J. Soil Sci.* 16:149–157.

Went, F.W., and N. Stark. 1968. The biological and mechanical role of soil fungi. *Proc. Nat. Acad. Sci.* 60:497–504.

Yoder, R.E. 1936. A direct method of aggregate analysis of soils and a study of the physical nature of erosion losses. *J. Am. Soc. Agron.* 28:337–351.

Zvyagintsev, D.G., A.F. Pertsovskaya, V.I. Duda, and D.I. Nikitin. 1969. Electron microscopic study of the adsorption of micro-organisms on soil and minerals. *Microbiology* 38:937–942.

# The Distinctive Properties of Andosols

### Koji Wada*

*Faculty of Agriculture, Kyushu University 46, Fukuoka 812, Japan.

© 1985 Springer-Verlag New York, Inc.
Advances in Soil Science, Volume 2.

# I. Introduction

Most soils derived from volcanic ash and pumice show distinctive properties that are not found in soils derived from other parent materials under the same vegetation and climate. Thorp and Smith (1949) first recognized them as a great soil group and gave them the tentative name Ando soils; *Ando* means dark soils in Japanese. They noted a wide distribution of these soils in deposits of volcanic ash not only in Japan but in other parts of the world. The Ando soils consist, primarily, of dark brown to black $A_1$ horizons, averaging about 30 cm thick, of fine crumb or granular structure with the organic content close to 8% on average and ranging up to 30% in the darkest members of the group. Some members of the group have distinct B horizons with more clay than A horizons, but the younger members are essentially AC soils. The soils have low exchangeable "bases" and occur in humid to perhumid climates with temperature efficiency ranging from cool mesothermal to tropical. Natural vegetation varies from place to place and includes broad-leaved and coniferous forest types often with an understory of bamboo. Thorp and Smith (1949) had some difficulty in deciding whether Ando soils should be placed in the zonal or intrazonal orders, because they did not know whether the dark color of the soils was caused by vegetation or by the soils' parent material.

The names Andosols and Andepts are used in the legend of FAO/ UNESCO *Soil Map of the World* (1974) and in *Soil Taxonomy (Soil Survey Staff*, 1975), respectively, for soils having a bulk density of the fine earth fraction of the soil of less than 0.85 g cm$^{-3}$ and an exchange complex dominated by amorphous material and/or having 60% or more vitric volcanic ash, cinders, or other vitric pyroclastic material in the silt, sand, and gravel fractions. This definition implies that every soil derived from volcanic ash and pumice is not an Andosol or Andept. On the other hand, they can be formed from parent materials other than volcanic ash and pumice.

Then, Smith* proposed elevating the suborder of Andepts in *Soil Taxonomy* to an order called Andisols and introducing soil moisture and temperature regimes to define suborders. Andisols were then defined as soils developing in volcanic ash, pumice, cinders, and other volcanic ejecta with an exchange complex that is dominated by X-ray amorphous compounds of Al, Si, and humus, or a matrix dominated by glass, and having one or more diagnostic horizons other than an ochric epipedon. Bulk densities are always comparatively low in most horizons, though the absolute values vary with the degree of weathering, the humidity of the soil climate, and in a very few with the degree of cementation by silica or other cements. The most diagnostic horizons are an umbric, or rarely a mollic,

*G.D. Smith, 1978. A preliminary proposal for reclassification of Andepts and some Andic subgroups. Unpublished letter. 20 pp.

epidedon and a cambic horizon, or an ochric epipedon and a cambric horizon.

Smith[1] listed several important properties common to Andisols. The amorphous clays have a low permanent charge and a high pH-dependent charge. Aluminum toxicity is rare. Phosphate fixation and water retention are high relative to most other soils with comparable textures. Percentages of carbon tend to be high relative to other mineral soils, but bulk densities tend to be low, and the weight of carbon per unit volume does not differ so greatly as the percentages.

The object of this article is to summarize the available information about the nature and properties of soils derived from volcanic ash and pumice, which contributes to the establishment of their proper management and use. The name Andosols is used for them for brevity, though some of them do not meet the definition of Andosols. The article essentially consists of three parts: the first part describes the formation of Andosols by weathering of volcanic ash mainly under temperate and humid conditions; the second, the humus and clay minerals that form in these processes and give distinctive properties to Andosols; and the third, the account and implication of distinctive properties of Andosols.

## II. Formation of Andosols

Zero time of soil formation, and hence the time for horizon differentiation, can easily be established in soils developed from volcanic ash on the stable surface. In Hokkardo, Japan, which has cool, moist summers, and snow in winter, Yamada (1968) derived the following figures for the period of horizon differentiation: 100 yrs for C or (A)C, 100 to 500 yrs for AC or A(B)C, 500 to 1000 yrs for A(B)C or ABC, and >1000 yrs for ABC. The ideal horizon differentiation however, frequently cannot be established owing to the shallowness of deposited ash and to successive showers of ash combined with rapid soil development.

A good example of Andosol development under a humid temperate climate is seen in the grassland soils of the Mt. Aso district in Kyushu, Japan (Wada and Aomine, 1973). In this district, the continuing activity of the central cones and the prevailing westerly winds result in deposition of the ashes showing a geographical variation. This is reflected in the geographical variation of the C content of the surface soil (figure 1). Ashes are deposited most frequently and thickly around the central cones, followed by the eastern and northeastern somma, and most thinly on the northern and western somma. The carbon content of the surface soil increases in this order.

Analytical data for the representative soils in three regions in the Mt. Aso district are given in Table 1. An exotic ash "Imogo," whose age was estimated to be 6000 to 6500 yrs (Machida and Arai, 1978), is seen at a

**Table 1.** Analytical Data for the Representative Andosols in the Mt. Aso District[a]

| Depth (cm) | C (%) | C/N | Clay[b] | Bulk density | PSC[c] | CEC (meq /100 g) | BS[d] (%) | pH (H$_2$O) |
|---|---|---|---|---|---|---|---|---|
| Site 220 (central cone) | | | | | | | | |
| 0–12 | 1.75 | 13 | 3 | 1.2 | – | 6 | 62 | 5.7 |
| 12–24 | 1.30 | 13 | 6 | 1.3 | 1150 | 6 | 53 | 6.0 |
| 24–35 | 2.23 | 12 | 12 | 0.9 | 1480 | 10 | 56 | 6.0 |
| 35–63 | 2.41 | 13 | 14 | 0.8 | 1480 | 13 | 58 | 6.2 |
| 63–68 | 1.26 | 13 | 12 | 0.95 | 1320 | 10 | 51 | 6.0 |
| 68–109 | 1.71 | 19 | 12 | 1.0 | 1240 | 12 | 46 | 6.1 |
| Site 217 (northeastern somma) | | | | | | | | |
| 0–16 | 6.9 | 12 | 16 | 0.65 | 2150 | 23 | 27 | 5.6 |
| 16–35 | 8.5 | 14 | 25 | 0.50 | 2320 | 31 | 23 | 5.6 |
| 35–50 | 14.8 | 18 | 38 | 0.35 | 2730 | 63 | 26 | 5.4 |

| | | | | | | | |
|---|---|---|---|---|---|---|---|
| 50–67 | 18.4 | 18 | 66 | 0.45 | 2650 | 74 | 18 | 5.6 |
| 67–73 | 6.1 | 13 | 63 | 0.40 | 2320 | 36 | 10 | 5.6 |
| 73–100 | 8.4 | 16 | 68 | 0.45 | 2480 | 37 | 10 | 5.6 |
| Site 206 (northern somma) | | | | | | | |
| 0–13 | 15.5 | 17 | 25 | 0.40 | 2320 | 36 | 17 | 5.1 |
| 13–38 | 21.7 | 23 | 50 | 0.30 | 2880 | 57 | 6 | 5.0 |
| 38–65 | 16.4 | 21 | 78 | 0.25 | 2800 | 46 | 4 | 5.4 |
| 65–95 | 19.0 | 28 | 75 | 0.25 | 2960 | 57 | 5 | 5.2 |
| 95–105 ("Imogo") | 3.5 | 15 | 49 | 0.25 | 2650 | 7 | – | 5.7 |

[a] Adapted from Wada and Aomine (1973), *Soil Science* 116:170–177. Copyright © 1973 by The Williams & Wilkins Co., Baltimore. Used with permission.

[b] The total amount of <2 μm fraction and material dissolved by Jeffrie's treatment (nascent hydrogen–potassium oxalate–oxalic acid) on the inorganic material basis.

[c] Phosphate sorption coefficient, i.e., phosphate adsorption from 2.5% ammonium phosphate aq., pH 7.0, measured using a soil solution ratio = 10 g/20 ml and reported as mg $P_2O_5$/100 g air-dry soil.

[d] Exchangeable "base" saturation.

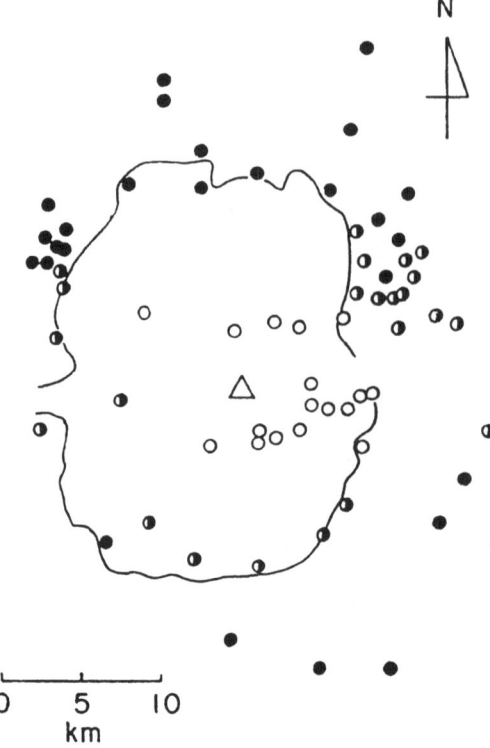

**Figure 1.** Geographical variation in the C content of surface soils in Mt. Aso district. Triangle = the central cone; open circle = C% <5; half-closed circle = C% 5 to 10; closed circle = C% >10. The line indicates approximately a summit line of the somma. From Wada and Aomine (1973), *Soil Science* 116:170–177. Copyright © 1973 by The Williams & Wilkins Co., Baltimore. Used with permission.

depth of 100 cm or less on the northern somma. The same ash is normally seen at a depth of 150 cm or more on the northeastern somma, but not even at the greatest depth under observation around the central cone. The differences in the soil characteristics between the three regions can be correlated with those in the age of the soils. The data explain what characteristics have been acquired by the soil at different stages of soil formation and indicate that Andosols can attain maturity within 2000 to 3000 yrs.

Column C (%) in Table 1 illustrates that humus accumulation constitutes one of the striking features of Andosols. The C content of each soil in the profiles indicates its residence time at or near the surface where the soil received a supply of organic matter from plant residue. The differences in the C content and its variation with depth among the three profiles again reflect the geographical variation in the deposition of volcanic ash. The sum total of humus accumulated for 6000 to 6500 yrs in both the surface and buried soils is calculated to amount to 1700 t ha$^{-1}$. Note that this humus accumulation has occurred under a warm, humid climate (mean summer and winter temperature, 20 to 40°C and 0 to 4°C; annual precipitation, 2000 to 2400 mm) and at well-drained sites on the gently rolling topography. The annual production of natural grass (major

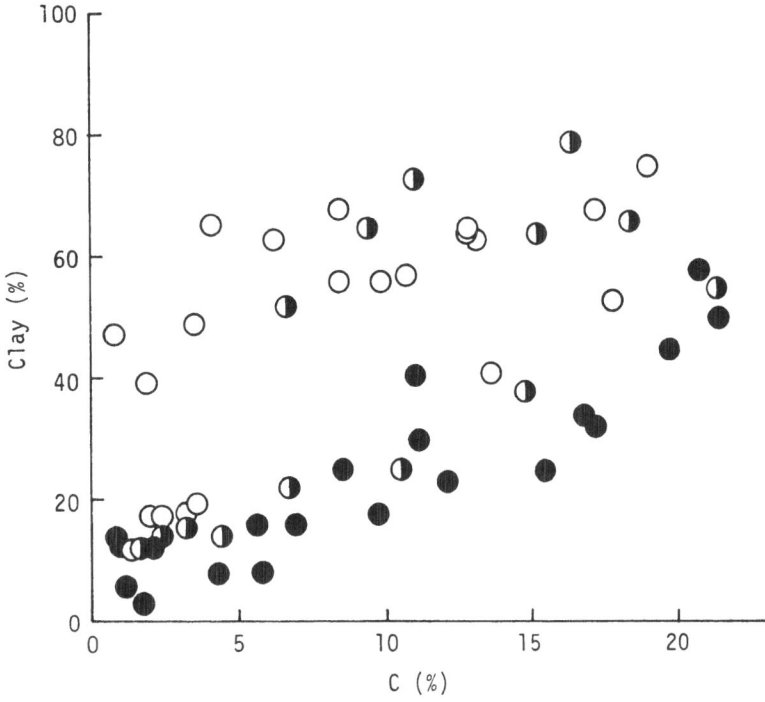

**Figure 2.** Relation between humus accumulation and clay formation for Andosols in Mt. Aso district. Closed circle = soil at depth of 0 to 35 cm; half-closed circle = soil at depth of about 35 to 70 cm; open circle = soil at depth of about 70 to 100 cm. Based on data from Wada and Aomine (1973) and S. Aomine and K. Wada, unpublished data.

species, *Miscanthus sinensis* and *Sasa*) was ranged from 3 to 10 t ha$^{-1}$ of dry matter (Wada and Aomine, 1973).

The amount of clay formed in the surface soil correlates with the amount of humus accumulated, and the subsoil has higher clay content than the surface soil at the same C content (Figure 2). The latter indicates that the clay formation proceeds after the burial of the soil. The high clay content of the subsoil at 60 cm or greater depth on the northeastern and northern somma also illustrates the rapid soil formation from volcanic ash.

Andosols have also formed in climates different from the humid temperate. In Alaska, Cryandepts are extensive and grade into Spodosols (Flach, 1969). They are defined as having a mean annual soil temperature of less than 8°C and a mean summer soil temperature of less than 15°C. In Hawaii, Andosols have a mollic epipedon in place of umbric epipedon (Eutrandepts) in the state's subhumid region, whereas they are very hydrous (Hydrandepts) in its perhumid region, most probably reflecting the changing mineralogy from allophane to amorphous hydroxides (Sherman and Swindale, 1964).

In South America, Wright (1964) classified Andosols into two groups, the high-latitude group of Chile and Argentina and the low-latitude group of Ecuador and Colombia, and stressed the importance of moisture regime that correlates with the minor variations in soil morphology and with land use properties. Though he considered that both groups of soils have adequate soil moisture to produce significant amounts of allophane, a recent examination of Chilean and Ecuadoran Andosols showed that there are important differences in this and other regards between the two groups of soils (Sixth International Soil Classification Workshop, 1984). In Kenya, Wielemaker and Wakatsuki (1984) reported soils with morphologies typical of Andosols but lacking certain other characteristics common to such soils.

## III. Humus and Its Accumulation

As illustrated in Table 1, the accumulation of humus is a striking feature of Andosols. Table 2 shows the C content found in the pairs of adjacent virgin and cropped Andosols from northern to southern Japan (Adachi, 1973). There is no particular difference in the content of humus in the virgin soils of the three districts, indicating the predominating influence of moisture over the temperature regime. However, humus decreases by cultivation, and this decrease is more marked in the south than the north.

**Table 2.** C Content and C/N Ratio of Andosols in Japan[a]

| District | Number of samples | Uncultivated soil | | Cultivated soil | |
|---|---|---|---|---|---|
| | | C (%) | C/N | C (%) | C/N |
| Northern | 9 | 1–19 | 8–21 | 5–16 | 14–20 |
| 6–12°C[b] | | 13.5 | 15.8 | 9.0 | 16.0 |
| 800–1800 mm[c] | | (±3.7) | (±3.6) | (±3.8) | (±2.4) |
| Central | 9 | 9–20 | 14–18 | 4–13 | 11–16 |
| 12–16°C | | 12.9 | 15.8 | 7.6 | 13.7 |
| 1200–2400 mm | | (±3.7) | (±1.5) | (±2.7) | (±2.1) |
| Southern | 8 | 7–20 | 13–21 | 3–11 | 13–21 |
| 16–17°C | | 14.5 | 16.5 | 7.3 | 15.5 |
| 1800–2500 mm | | (±5.5) | (±2.7) | (±2.8) | (±3.1) |

[a]Adapted from Adachi (1973); the figures indicate, from top to bottom, the range, mean, and standard deviation of the measured values.
[b]Average annual temperature.
[c]Average annual rainfall.

In New Zealand, most Andosols are formed under a mild, humid climate (mean summer and winter temperature, 15 to 19°C and 6 to 9°C; annual rainfall, 1100 to 2500 mm). They have humic topsoils, but the $A_1$ horizon is not very thick ($\leq 15$ cm) and its carbon content ranges from 3.5 to 11.6% (average, 9.3%) in 15 representative soil profiles (Gibbs, 1968). Table 3 summarizes the relations between climate, vegetation, and humus accumulation observed for different groups of Andosols in South America (Wright, 1964). The humus content ranges widely and the relations are rather complicated, but the following tendencies are noted: (1) the humus content is high in the continuously humid zone and even higher in the perhumid zone for the soils of Chile and Argentina; and (2) it decreases slightly with increasing temperature for the soils of Ecuador and Colombia. In the subhumid to perhumid regions of Hawaii, the carbon content of soils derived from volcanic ash ranges from 5 to 15% in the A horizon and 0.5 to

**Table 3.** The Relation Between Climate, Vegetation, and C Content of Andosols in South America[a]

| Soil group number | Climate data[b] | Vegetation | C content (%) |
|---|---|---|---|
| (A) High-latitude group of Chile and Argentina | | | |
| 1 | 800–1000 mm 25–10°C | Evergreen and semi-deciduous | 5–7 |
| 2 | 1000–2500 mm 25–9°C | Deciduous with evergreen | 9–16 |
| 3 | 2000–2500 mm 25–9°C | Evergreen (*Nothofagus*) | 12–16 |
| 4 | 2000 mm 17–7°C | *Nothofagus* | 6–9 |
| 5 | 3000–5000 mm 17–7°C | *Nothofagus* | 15–17 |
| (B) Low-latitude group of Ecuador and Colombia | | | |
| 8 | 3000–5000 mm 8–10°C | Stipa grass and subalpine perennials | 7–13 |
| 9 | 800–1200 mm 12–16°C | (Forest) | 3–4 |
| 10 | 2000–3500 mm 15–20°C | Montane rain forest | 13–16 |
| 11 | 1500–3000 mm 18–21°C | Rainforest | 7–10 |
| 12 | 2500–3500 mm 21–24°C | Rainforest | 5–10 |

[a]Adapted from Wright (1964).
[b]Top, average annual rainfall; bottom, mean summer–winter temperature for (A) group of soils and mean annual temperature (fairly equable) for (B) group of soils.

7% in the B horizon, and the organic content is highest in the subhumid region (Sherman and Swindale, 1964).

The humus in Andosols is characterized by a wide C/N ratio, and commonly there are values over 13 (Tables 1 and 2). Similar values were reported for Andosols in New Zealand, except those called Red Loams and Brown Loams, which had narrow C/N ratios of about 11 (Gibbs, 1968). The C/N ratios ranging from 12.5 to 16.2 were found in Chile (Aomine, 1972). On the other hand, a narrow C/N ratio of 10 to 11 was found in Andosols formed in the subhumid regions of Hawaii, and a slightly higher C/N ratio of 14 in the perhumid region (Sherman and Swindale, 1964).

The reported humic acid carbon ($C_h$)/fulvic acid carbon ($C_f$) ratios of Japanese Andosols are 0.8 to 1.8 (Tokudome and Kanno, 1968), 0.4 to 1.4 (Adachi, 1973), and 0.6 to 3.0 (Kobo and Oba, 1974a). Arai (1983) found that the $C_h/C_f$ ratios of Andosols in the humid (annual rainfall, 1800 to 3000 mm) and subhumid to dry (annual rainfall, 600 to 2300 mm) regions of Hawaii are lower and higher than 0.5, respectively. Tan (1964) noted a similar difference between the Lowland ($\leq 0.2$) and Highland ($\geq 0.5$) Andosols in Java. There were broad parallels between the increase of humus content, the C/N ratio, and the $C_h/C_f$ ratio of the color deepening of humic acid for Japanese Andosols (Table 4). These parallels were interpreted as indicating that humification proceeds with humus accumulation. Kobo and Oba (1974a) also reported that the amount, the C/N ratio, and the form of humus are not affected by the petrographic nature of parent volcanic ash.

Well-humified humic acids in Andosols do not have particularly high molecular weights. Table 5 shows the result of fractionation by gel filtration for humic acids that were extracted from an Andosol and an alluvial soil (Inoko and Tamai, 1976). If the high C/N ratio and the relative coloration of solution are taken as indications of humification, well-humified humic

**Table 4.** A Summary of Analytical Data on Humus in Japanese Andosols[a]

| Humus content (%) | Number of samples | C/N ratio | HE/HT[b] (%) | $C_h/C_f$ ratio | $\Delta \log k$[c] |
|---|---|---|---|---|---|
| <10 | 3 | 16.1 | 68 | 0.98 | 0.619 |
| 10–20 | 20 | 16.3 | 66 | 1.22 | 0.561 |
| 20–30 | 17 | 19.0 | 72 | 1.65 | 0.530 |
| >30 | 6 | 20.1 | 73 | 2.07 | 0.515 |

[a]Adapted from Kobo and Oba (1974a), *Journal of Science of Soil Manure* 45:227–233. Copyright © 1974 by the Japanese Society of Soil Science and Plant Nutrition. Used with permission.

[b]Percent of humus extracted with hot 0.1 $M$ NaOH in total humus.

[c]$\log k_{400} - \log k_{600}$, where $k_{400}$ and $k_{600}$ are absorbance of humic acid solution at 400 and 600 nm, respectively.

Table 5. Humic Acids Extracted from an Andosol and an Alluvial Paddy Soil in Different Sephadex Molecular Weights[a]

| Sephadex molecular weight | Andosol | | | Alluvial soil | | |
|---|---|---|---|---|---|---|
| | Distribution (%) | C/N ratio | Relative[b] coloration | Distribution (%) | C/N ratio | Relative[b] coloration |
| >200,000 | 15 | 10.5 | 11.5 | 9 | 9.9 | 7.3 |
| 200,000–50,000 | 17 | 9.6–10.2 | 16.7–18.1 | 17 | 9.8–11.6 | 9.8–14.1 |
| 50,000–10,000 | 27 | 11.5 | 21.8 | 27 | 11.1 | 17.3 |
| 10,000–1000 | 37 | 11.0–14.2 | 27.0–37.7 | 43 | 12.5–13.2 | 20.4–28.1 |
| <1000 | 5 | 17.8 | 18.6 | 5 | 19.6 | 26.7 |

[a] Adapted from Inoko and Tamai (1976). Humic acid was extracted by 0.1 $M$ $NA_4P_2O_7$ (pH 7.0) at 28°C. The humus content and C/N ratio were 19.0% and 16.1 for the Andosol and 5.0% and 11.6 for the alluvial soil, respectively.
[b] $-\log T_{370}/C$ mg/ml, where $T_{370}$ is the transmittance of humic acid solution at 370 nm.

acids have rather lower molecular weights. This observation is supported
by infrared spectroscopy; the high molecular weight fractions showed
stronger aliphatic C–H stretching absorption and weaker $COO^-$ absorption than did the low molecular weight fractions. Similar observations were
made by Yonebayashi (1976).

The large amount of humus accumulation in Andosols has often been
attributed to the greater annual supply of plant residues under grass than
under forest. Table 6 illustrates the effect of previous vegetation on the
humus in Andosols carrying forest (*Chamaecyparis* or *Fagus*) for about 50
yrs. It is evident that the introduction of grass not only increased the
content of humus but affected the nature of humus as illustrated in the color
and the C/N ratio of the soil.

The differences in microflora between soils derived and not derived from
volcanic ash have been studied in Japan in relation to the accumulation of
humus. As illustrated in Table 7, there were no significant differences in the
number of bacteria and fungi between the two groups of soils, whereas
actinomycetes and anaerobic bacteria were more abundant in the soils
derived from volcanic ash, irrespective of whether the soils are cultivated or
not (Ishizawa and Toyoda, 1964). Among the actinomycetes, chromogenic
species producing strong brown pigments were abundant, particularly
under natural vegetation (Araragi and Ishizawa, 1972). Cultivation
affected the microflora in both the volcanic ash soils and non-volcanic ash
soils, but there was no particular difference between the two groups of
soils.

The mechanism of humus accumulation in Andosols has attracted the
attention of many investigators. As reviewed by Wada and Higashi (1976),
earlier, the clay–humus interaction—specifically the allophane–humus
interaction—was suggested to be important in view of probable protection
of humus against the attack of microorganisms. The preformation of
allophane and related minerals in Andosols is implicitly assumed when their
importance to the accumulation of humus is inferred. The absence or near
absence of allophane and related minerals in some Andosol $A_1$ and buried
$A_1$ horizons in which considerable amounts of humus accumulated was,
however, found by Tokashiki and Wada (1975) and subsequently by a
number of investigators. On the other hand, Kato (1970) pointed out that
sesquioxides made soluble by $H_2O_2$ and dithionite–citrate treatment, rather
than allophane, are important to the accumulation of humus in some soils
derived from volcanic ash. Kobo and Oba (1974b) found that many
Andosols show no apparent correlation between the total C and the Al
extracted by Tamm's A reagent (ammonium oxalate–oxalic acid mixture)
that dissolves non-crystalline clay constituents.

The forms of inorganic soil constituents complexed with humus were
studied by extracting the soil with solutions of pyrophosphate and
dithionite–citrate and by determining the amounts of C, Al, and Fe
extracted. Sodium pyrophosphate solutions were found to be effective as

**Table 6.** Effects of Previous Vegetation on Humus in the Two Sets of Forest Soils ($A_1$ Horizon) Derived from Volcanic Ash[a]

| Soil | Vegetation | | Soil color | C(%) | C/N ratio | $C_h/C_f$ ratio | $\Delta \log k$[b] |
|---|---|---|---|---|---|---|---|
| | Present | Previous | | | | | |
| Fuji-1 | *Chamaecyparis obtusa* (48 yrs) | *Fagus crenata* | 7.5YR 2/2 | 20.4 | 12.9 | 0.60 | 0.626 |
| Fuji-2 | *Chamaecyparis obtusa* (53 yrs) | Natural grass | 7.5YR 1.7/1 | 27.1 | 16.8 | 1.64 | 0.518 |
| Amagi-1 | *Fagus crenata* (64 yrs) | *Fagus crenata* | 7.5YR 2/2 | 17.3 | 17.6 | 0.51 | 0.604 |
| Amagi-2 | *Chamaecyparis obtusa* (45 yrs) | Natural grass | 7.5YR 1.7/1 | 20.1 | 22.3 | 1.22 | 0.474 |

[a] Adapted from Yagi et al. (1983).
[b] $\log k_{400} - \log k_{600}$, where $k_{400}$ and $k_{600}$ are absorbance of humic acid solution at 400 and 600 nm, respectively. Humic acid was extracted with hot 0.1 $M$ NaOH.

**Table 7.** Numbers of Microorganisms in Soils Derived and Not Derived from Volcanic Ash in Japan[a]

| | Average numbers of microorganisms g$^{-1}$ dry soil[b] | | | |
| | Natural vegetation | | Cultivated | |
| Microorganisms | Volcanic ash soils (n=23)[c] | Other soils (n=17) | Volcanic ash soils (n=31) | Other soils (n=27) |
| --- | --- | --- | --- | --- |
| Bacteria | $8.4 \times 10^6$ | $12.1 \times 10^6$ | $18.1 \times 10^6$ | $21.8 \times 10^6$ |
| | $(4.5 \times 10^6)$ | $(7.9 \times 10^6)$ | $(11.6 \times 10^6)$ | $(11.4 \times 10^6)$ |
| Actinomycetes | $4.7 \times 10^6$ | $2.3 \times 10^6$ | $8.0 \times 10^6$ | $4.8 \times 10^6$ |
| | $(2.5 \times 10^6)$ | $(1.2 \times 10^6)$ | $(3.3 \times 10^6)$ | $(2.5 \times 10^6)$ |
| Anaerobic | $14.8 \times 10^4$ | $8.3 \times 10^4$ | $23.4 \times 10^4$ | $14.7 \times 10^4$ |
| bacteria | $(8.0 \times 10^4)$ | $(6.4 \times 10^4)$ | $(12.9 \times 10^4)$ | $(9.5 \times 10^4)$ |
| Fungi | $11.8 \times 10^3$ | $13.5 \times 10^3$ | $24.1 \times 10^3$ | $23.1 \times 10^3$ |
| | $(9.0 \times 10^3)$ | $(6.1 \times 10^3)$ | $(9.3 \times 10^3)$ | $(9.6 \times 10^3)$ |

[a]Adapted from Ishizawa and Toyoda (1964).
[b]Number in parentheses shows standard deviation.
[c]n = number of sample.

extractants of Al and Fe from their organic complexes (Bascomb, 1968), whereas dithionite–citrate was found to be effective as an extractant of Al and Fe from "free" Fe oxides as well as their organic complexes (McKeague and Day, 1966). Figures 3 and 4 show the results of such extraction studies on Japanese Andosols of different origins and ages (Wada and Higashi, 1976). There is an approximately linear relation between the C extracted by pyrophosphate ($C_{pyr}$) and the Al and Fe extracted by the same reagent [$(Al + Fe)_{pyr}$] for all soils (Figure 3), which suggests that $Al_{pyr}$ and $Fe_{pyr}$ form a metal humus complex with a stoichiometric ratio of metal to carbon ranging from 0.17 to 0.23.

In Figure 4, the residual C ($C_{res}$ = total C − $C_{pyr}$) is plotted against the Al and Fe , which were extracted by dithionite–citrate but not by pyrophosphate [$(Al + Fe)_{dit-pyr}$]. The soils are grouped into four zones. Zone I includes all the soils of age group 1 (<1000 yrs) containing very little $(Al + Fe)_{dit-pyr}$ in spite of wide variation of $C_{res}$. Zone II includes all $A_1$ horizons and buried $A_1$ horizons of age group 2 (<2500 yrs), which contain high amounts of $C_{res}$ but rather small amounts of $(Al + Fe)_{dit-pyr}$. Zone III contains the buried $A_1$ horizons of age group 3 (2500 to 5000 yrs) with an approximately linear relation between $C_{res}$ and $(Al + Fe)_{dit-pyr}$. Zone IV contains the B and buried B horizons and the oldest buried $A_1$ horizons.

Figures 3 and 4 suggest that the youngest humus that has a very low complexing ability for Al and Fe (and is not extracted into pyrophosphate) evolves with time or soil formation into forms that complex Al and Fe (and is extracted into pyrophosphate). When the supply of organic matter is

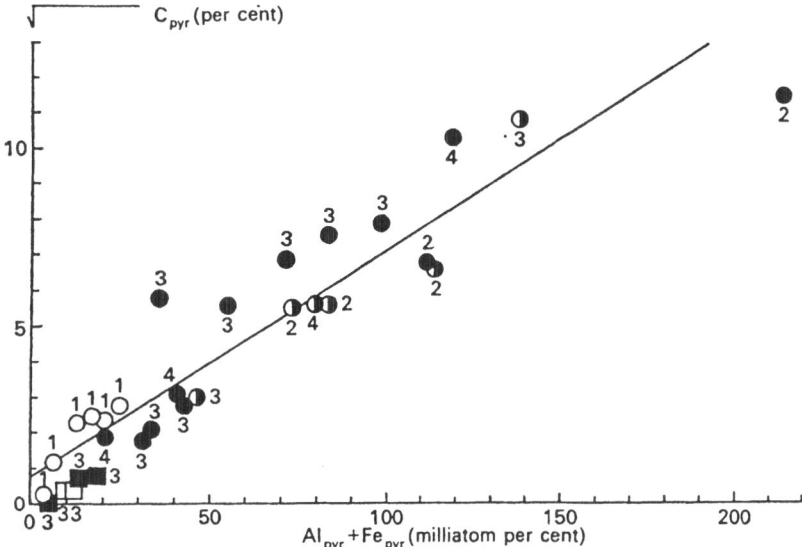

**Figure 3.** Relation between pyrophosphate-extractable C and (Al+Fe) in Andosols. Open circle = $A_1$ (young); half-closed circle = $A_1$ (older); closed circle = $A_1b$; open square = (B); closed square = (B)b. The numbers indicate the age group (see text) to which each soil belongs. From Wada and Higashi (1976), *Journal of Soil Science* 27:357–368. Copyright © 1976 by the British Society of Soil Science. Used with permission.

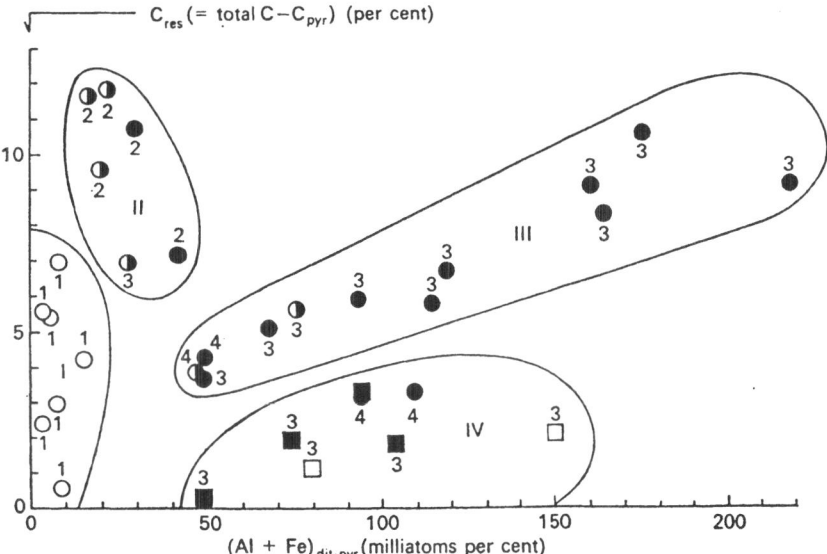

**Figure 4.** Relation between total minus pyrophosphate-extractable C and dithionite–citrate minus pyrophosphate-extractable (Al+Fe) in Andosols. For legend see Figure 3. From Wada and Higashi (1976), *Journal of Soil Science* 27:357–368. Copyright © 1976 by the British Society of Soil Science. Used with permission.

limited by burial of the soil, the humus evolves further into forms that are not extracted by pyrophosphate, perhaps by reaction with additional Al and Fe released by continued weathering of volcanic ash. Some of such Al and Fe may be present as hydrous oxides and allophanelike constituents (and is extracted by dithionite–citrate) and some as allophane and imogolite. A similar reaction would take place with the young humus that migrated into the color B horizons.

Most probably, parts of Al and Fe that complex with humus are present as polymer hydroxyions. A trace or only a small amount of Al was extracted by 1 $M$ KCl from Andosols containing a large amount of Al that complexes with humus, unless they contain 2:1 layer silicates (Shoji and Ono, 1978; Saigusa *et al.*, 1980). The correlation between phosphate adsorption and humus content, as illustrated in Figure 13, was interpreted in terms of the reaction of such polymer hydroxy-Al and -Fe ions with phosphate (see Table 13). A blocking of ionized carboxyl groups of humus by hydrox-Al and -Fe ions was also inferred from the study on the charge characteristic of Andosol $A_1$ horizons (see Table 10).

The reactions between the inorganic constituents and humus result in the formation of clay-, silt- and sand-sized aggregates in Andosols that are stable even to sonic oscillation (Higashi and Wada, 1977). Both the C content of the clay-sized separate and the difference in the C content between the clay and the silt- or sand-sized separates are higher in the younger soils (<2500 yrs) than in the older soils. This demonstrates an important role of Al- and Fe-humus complexes in the aggregate formation and the growth of the aggregates with soil formation.

## IV. Clay Minerals Particular to Andosols

The formation of clay minerals is the most prominent process in the formation of soil from igneous rocks. Andosols contain unique clay minerals such as allophane, allophanelike constituents, imogolite, and opaline silica. They contribute to the development of the distinctive properties of Andosols. The following sections briefly describe these minerals (Wada, 1977, 1980).

### A. Allophane

Allophane was described as an amorphous material having little or no structural organization, but high-resolution electron microscopy has indicated that allophane consists of "hollow spherules" with diameters of 3.5 to 5 nm (Figure 5). The variability of allophane's shape and size results from aggregation of these spherules with themselves and with other clay constituents. Figure 6 illustrates a probable structure of an allophane spherule proposed on the basis of the data on morphology, chemical

**Figure 5.** An electron micrograph of allophane. From Wada and Wada (1977), *Clay Minerals* 12:289–298. Copyright © 1977 by the Mineralogical Society, London. Used with permission.

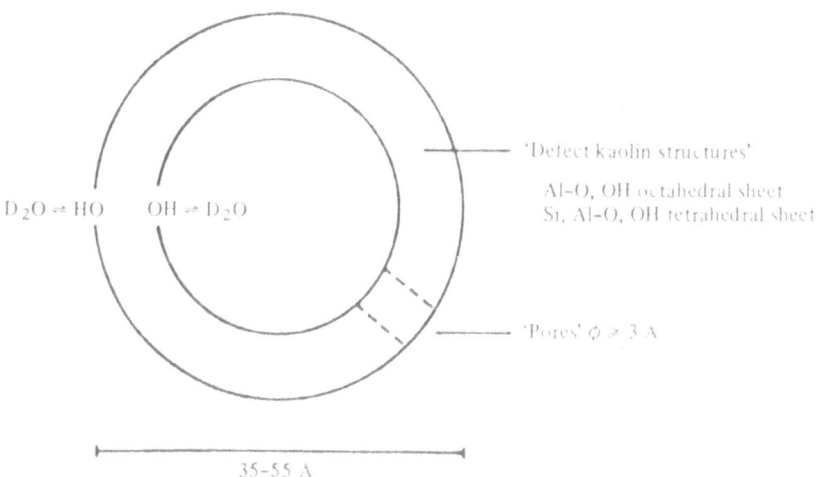

**Figure 6.** A schematic presentation of the structure of an allophane spherule. From Wada (1980), in *Soils with Variable Charge*, pp. 87–107. Copyright © 1980 by the New Zealand Society of Soil Science. Used with permission.

composition ($SiO_2/Al_2O_3$ molar ratio, 1.0 to 2.0; $H_2O$ (+)/$Al_2O_3$ molar ratio, 2.5 to 3.0), Al-coordination (6- and 4-fold), density (2.75 g cm$^{-3}$), and a rapid and complete OH–OD exchange. The feature of the infrared spectrum in the region between 400 and 800 cm$^{-1}$ supports the defect imogolite rather than the defect kaolin structure of the allophane spherule. The calculated and measured specific surface areas of allophane are 1050 m$^2$g$^{-1}$ and 700 to 1100 m$^2$g$^{-1}$, respectively. Allophanes are dissolved by treatment with hot 0.5 $M$ NaOH or 0.15 to 0.2 $M$ oxalate–oxalic acid (pH 3.0 to 3.5). The analysis of Si and Al dissolved and the infrared spectrum of the dissolved constituents recorded by a difference method are useful for the determination of allophane.

### B. Allophanelike Constituents

Unlike allophane, allophanelike constituents are dissolved by treatment with dithionite–citrate and 2% $Na_2CO_3$ solution. They showed infrared spectra similar to allophane, imogolite, and synthesized hydroxyaluminosilicate cations. Allophanelike constituents coexist with or may be a part of allophane. They have a $SiO_2/Al_2O_3$ molar ratio lower than that of allophane and show a higher Al reactivity.

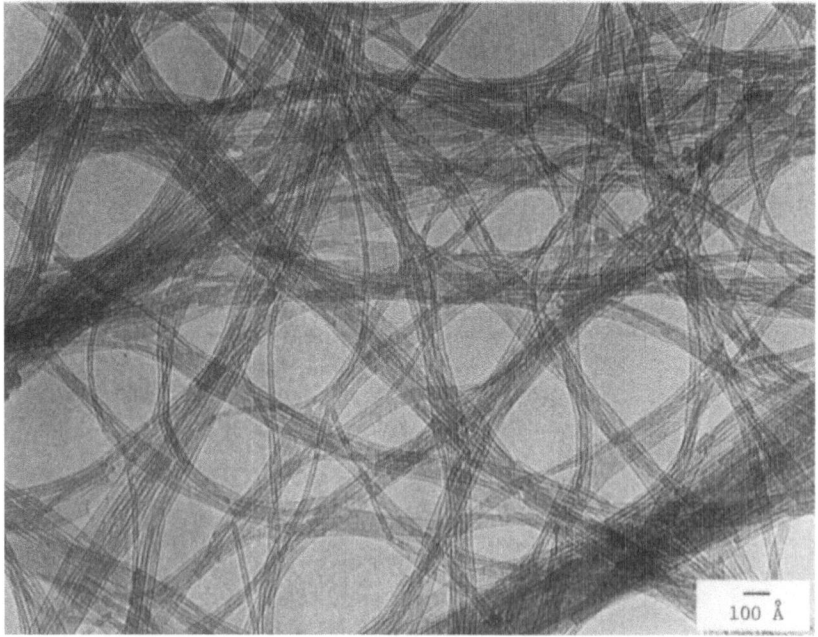

100 Å

**Figure 7.** An electron micrograph of imogolite. From Wada *et al.* (1972), *Clays and Clay Minerals* 20:375–380. Copyright © 1972 by The Clay Mineral Society. Used with permission.

## C. Imogolite

Imogolite was first described in a soil derived from volcanic ash "Imogo" (Table 1). It appears as smooth and curved threads with varying diameters from 10 to 30 nm and extending several μm in length. The threads consist of finer tube units with inner and external diameters of about 1.0 and 2.0 nm, respectively (Figure 7). The best empirical formula for imogolite is $1.1SiO_2 \cdot Al_2O_3 \cdot 2.3-2.8H_2O(+)$. A structure for the tube unit of imogolite is derived on the basis of the data on electron diffraction, morphology, chemical composition, Al-coordination (6-fold), density (2.6 to 2.75 g $cm^{-3}$), and a rapid and complete OH–OD exchange. As illustrated in Figure 8, the structure is based on a curled gibbsite sheet and contains Si–O–Al bonds but not Si–O–Si bonds. Imogolite is said to be paracrystalline, because some randomness is involved in the alignment of the tube units. The calculated and measured specific surface areas are 1025 $m^2g^{-1}$ and 900 to 1000 $m^2g^{-1}$, respectively. Imogolite is dissolved by treatment with hot 0.5 $M$ NaOH or oxalate–oxalic acid. It gives a characteristic X-ray diffraction pattern, infrared spectrum, and differential thermal analysis curve. Imogolite coexists with allophane in many Andosols, and even a trace amount of imogolite can be identified using an electron microscope.

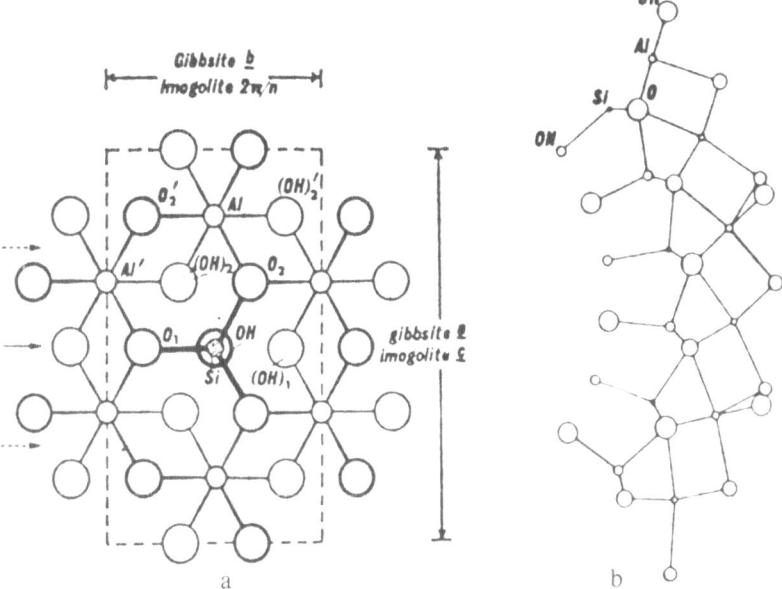

**Figure 8. a:** Posulated structure unit of imogolite in relation to that of gibbsite. The Si–OH groups that would lie at the cell corners in imogolite are omitted from the diagram. **b:** Curling of the modified gibbsite sheet induced by contraction of one surface to accommodate $SiO_3OH$ tetrahedra; projection along the imogolite c axis. From Cradwick *et al.* (1972). Reprinted by permission from *Nature Physical Science*, Vol. 240. Copyright © 1972 Macmillan Journals Limited.

## D. Opaline Silica

Opaline silica appears as laminar particles with diameters of 0.2 to 5 μm and with no particular structure as found in diatoms. The particles are commonly circular and elliptical and extremely thin. Opaline silica is soluble in hot 0.5 $M$ NaOH but not in oxalate–oxalic acid and gives the infrared spectrum characteristic of silica.

## E. Halloysite

In Andosols, halloysite appears in hydrated and dehydrated forms and typically as spherical particles with diameters of 0.04 to 1 μm or curled flakes. The spherical particles are composed of roughly concentric bands of halloysite layers. Peculiar forms (Kirkman, 1977; Wada and Mizota, 1982) and incipient forms (Wada et al., 1982) of halloysites have also been found in weathered pumices and volcanic ashes.

# V. Formation and Transformation of Clay Minerals

Recent mineralogical and chemical analyses of Andosols have indicated that (1) their clay mineral composition varies depending on the stage of soil formation, the horizon, the petrological nature of the parent volcanic ash, the thickness of overburden ash deposits, and probably other factors; and (2) the formation and transformation of clay minerals by weathering of volcanic ash and pumices are very much affected by accumulation of humus that forms complexes with Al and Fe and with some clay minerals (Wada, 1980). Table 8 summarizes the major soil components formed in the various horizons of Andosols at different stages of soil formation. Early, middle, and late stages roughly correspond to 0 to 2500, 2500 to 7500, and >7500 yrs after deposition of volcanic ash under humid, temperate climatic conditions.

## A. Allophane and Imogolite

As illustrated in Table 8, $A_1$ horizons, particularly at the early stage of soil formation from volcanic ash, contain opaline silica, humus, Al- and Fe-humus complexes, but not allophane and imogolite. On the other hand, (B) horizons, and $A_1$ and (B) horizons, buried by subsequent ashfalls, and to which little or no organic matter has been supplied, contain allophane and imogolite. This observation suggests that the humus formed in the soil is strongly bound to Al released from volcanic ash by weathering, and thereby limits the coprecipitation of Si and Al and hence the formation of allophane and imogolite. An example of pertinent data is shown in Figure 9, where the C content of soil samples, with and without allophane and imogolite, is plotted against the amount of Al as $Al_2O_3$ dissolved from the

**Table 8.** Major Soil Components Formed by Pedogenesis in Andosols under a Humid Temperate Climate and Natural Grass Vegetation[a]

| Horizon | Stage of pedogenesis | | |
|---|---|---|---|
| | Early | Middle | Late |
| Surface<br>A₁ | Humus<br>Al- and Fe-humus<br>O.S. | Al- and Fe-humus<br>A'- and A-humus | Al- and Fe-humus |
| Subsurface<br>B and Bb | A', A, Im<br>A', A, Im | A', A, Im<br>A', A, Im | Ht, Gb<br>Ht, Gb |
| Ab | Humus<br>Al- and Fe-humus | Al- and Fe-humus<br>A'-, A- and Im-humus<br>A', A, Im | Al- and Fe-humus |
| Surface and subsurface | Sm, Vt, M<br>Fe-ox | Sm-Ch, Vt-Ch, M<br>Fe-ox | Sm-Ch, Vt-Ch, Ch, M<br>Fe-ox |

[a]Adapted from Wada (1980), in *Soils with Variable Charge*, pp. 87–107. Copyright © 1980 by the New Zealand Society of Soil Science. Used with permission. Abbreviations: A = allophane; A' = allophanelike constituents; Ch = chlorite; Fe-ox = iron oxide minerals; Gb = gibbsite; Ht = halloysite; Im = imogolite; M = mica; O.S. = opaline silica; Sm = smectite; Sm-Ch = smectite-chlorite intergrades; Vt = vermiculite; Vt-Ch = vermiculite-chlorite intergrades.

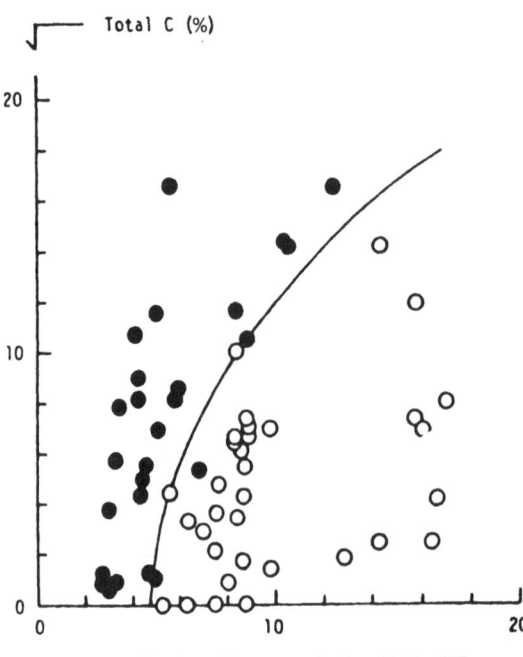

**Figure 9.** Relation between the humus accumulation and the formation of allophane (A) and imogolite (Im) in Andosol. Closed circle = $A_1$, Ap, and $A_1b$ without A and Im; open circle = Ap, B, $A_1b$, and Bb with A and Im. From Wada (1980), in *Soils with Variable Charge*, pp. 87–107. Copyright © 1980 by the New Zealand Society of Soil Science. Used with permission.

clay fraction by dithionite–citrate. The difference in the distribution of the samples, with and without allophane and imogolite, suggests that their formation is retarded when the accumulation rate of organic matter is high relative to the release rate of Al from volcanic ash. This supposition seems to be supported by subsequent observations that allophane and imogolite were found more frequently and in larger amounts in the Ap than in the $A_1$ horizons (Mizota and Wada, 1980; Mizota *et al.*, 1982) and under forest than under grass (Shoji *et al.*, 1982).

Allophane, allophanelike constituents, and imogolite usually coexist typically in the B horizons of Andosols (Table 8), provided the depositional overburden of volcanic ejecta is relatively thin (1 to 2 m). The leaching of bases and Si prevails in the formation of these minerals, though the pH levels over which allophane and imogolite are usually found are in the range from 5.0 to 7.0. Shoji *et al.* (1982) reported cases where the formation of allophane and imogolite from the same 1000 to 2000-yrs-old ashes is controlled by soil acidity. Allophane and imogolite are formed in the soils (pH >5.0) under 900 to 1200 mm rainfall and in *Castanea–Quercus* forests, but not in the soils (pH <4.8) under 1200 to 1800 mm or more rainfall and in *Abies* or *Fagus* forests. The latter soils show morphological features of Spodosol. Intergrading Andosols into Brown Podzolic soils was also found in southern Chile under rainfall in excess of 3000 mm and under coniferous or *Nothofagus* forest (Wright, 1964).

Allophane and imogolite have been found in soils other than Andosols. Brydon and Shimoda (1972) found allophane in Spodosols in Canada, and Tait *et al.* (1978) found imogolite in Podzols and Brown Forest soils on tills derived from granite, gneiss, and schist in Scotland. Subsequently allophane and/or imogolite have been found in a number of Spodosols, though not all examined Spodosols, and only in the B horizon but not in either the $A_1$ or $A_2$ horizon, as reviewed by Farmer (1982). Farmer *et al.* (1980) proposed that "proto-imogolite," a soluble aluminum silicate, rather than the soluble fulvate complexes is the predominant mobile form in which Al is transported from the A horizons to the $B_2$ and lower horizons. It is possible that allophane and imogolite are formed in the spodic B horizon, because the environmental conditions for the mineral formation may be similar in the subsurface horizons of some Andosols and Spodosols. The formation of "proto-imogolite" in the A horizons and its downward movement are, however, questioned from the observed effect of humus on the formation of allophane and imogolite in Andosols and from probable adsorption of "proto-imogolite" as polymer cations on coexisting negatively charged 2:1 layer silicates, respectively.

Imogolite (Farmer *et al.*, 1977; Farmer and Fraser, 1979) and allophane (Wada *et al.*, 1979) were synthesized by heating solutions containing dilute ($10^{-3}$ $M$) solutions of orthosilicic acid and $AlCl_3$ and varying amounts of NaOH at 95 to 100°C. The formation of imogolite was favored in the solutions containing Si and Al with the $SiO_2/Al_2O_3$ molar ratios at 1.0 and 2.0 and with pH 4.3 to 5.0. Allophanes are formed less susceptibly to the changes in the $SiO_2/Al_2O_3$ ratio (1.0 to 8.0) and pH (4.5 to 6.3) of the solution. These observations are useful in assessing the conditions for formation of these minerals in the natural environment.

### B. Opaline Silica

Opaline silica predominates in the $A_1$ horizon of young Andosols (<500 yrs old), where humus accumulation is active and the formation of allophane and imogolite is inhibited (Table 8). The actual mechanism of silica concentration and precipitation is not known. Wada and Nagasato (1983) obtained opaline silica by freeze-drying the $10^{-3}$ $M$ orthosilicic acid solution and found that the presence of Al inhibits this reaction. Opaline silica is rarely present in the B horizon and disappears gradually under leaching conditions, though its persistence was noted in soils 4000 to 7000 yrs old (Shoji and Masui, 1971).

### C. Non-crystalline Aluminum and Iron Oxides and Hydroxides

Iron oxides, mostly non-crystalline species, are common constituents of Andosols. Unlike Al, the stability of Fe in oxides is greater than that in humus complexes (Wada and Higashi, 1976). In tropical regions of high rainfall (>2500 mm), non-crystalline Al and Fe hydroxides rather than allophane and imogolite appear as major clay constituents. A fibrous form of poorly crystalline goethite was reported in two Andosols (Nakai and Yoshinaga, 1980).

## D. Halloysite

The relatively short life of allophane in a leaching environment under warm, humid climates is indicated from its transformation to halloysite in old and buried soils (Table 8). The formation of halloysite is usually favored by a thick depositional overburden of volcanic ejecta that produces a silica-rich environment. The age of the ash deposits containing halloysite ranges from about 10,000 to 250,000 yrs. Spherical halloysite particles grow with time, and their particle diameter ranges from 0.08 to 1.0 μm (Wada, 1980).

The effect of rainfall on the formation of allophane and halloysite was noted by Orbell *et al.* (1981) in weathering of 20,000 and 42,000-yr-old volcanic ash in New Zealand. The amount of halloysite in the two ash deposits increased with decreasing rainfall at the site from 2600 to 1200 mm. The dominance of halloysite in volcanic ash-derived soils in the Kingdom of Tonga (Claridge, 1981) may also reflect the effect of low rainfall.

The effect of rainfall on the formation of halloysite from volcanic ash in tropical regions, the Antilles, Ecuador, and Nicaragua, was noted by Colmet-Daage (1969). Halloysite was not found in regions of high rainfall (>2500 mm) but found in various forms and proportions in moderate rainfall regions (1500 to 2500 mm) with a pronounced dry season. In some intermediate regions, incipient forms of halloysite with amorphous substances appeared in young, sandy soils.

The formation of halloysite and "siliceous Fe oxides," both poorly ordered, were found in young Kenyan soils from trachytic volcanic ash under rainfall of 900 to 1300 mm (Wielemaker and Wakatsuki, 1984). The formation of incipient halloysites and probably their poorly crystalline precursors was found in young Ecuadoran soils derived from volcanic ash (K. Wada and Y. Kakuto, unpublished observations). These Kenyan and Ecuadoran soils containing poorly crystalline halloysites and their possible precursors have chemical and physical properties different from Andosols containing allophane and allophanelike constituents. A similar incipient form of halloysite was also found in two Japanese Andosols used as paddy that provides a silica-rich environment (Wada *et al.*, 1982).

## E. 2:1 and 2:1:1 Layer Silicates and Their Intergrades

Besides allophane, imogolite, opaline silica, and halloysite, more or less 2:1 and 2:1:1 layer silicates, and their intergrades and mixed-layer minerals, are usually present in Andosols. At the early stage of weathering, smectite, vermiculite, and mica are present, but the hydroxy-Al interlayering occurs in the former two minerals with advanced weathering (Table 8). The amount and distribution of these layer silicates show regional differences; relatively small amounts of the layer silicates were found in most soils derived from volcanic ash in Oregon, New Zealand, Chile, Ecuador, and Kenya. In Oregon, some Andosols contain a relatively high content of the layer silicates. It was ascribed to the admixture of the layer silicates present in the wall rock of a crator into volcanic ash during the eruption or to the incorporation of those from underlying paleosols (Dudas and Harward, 1975). In New Zealand, soils derived from "tephric loess" (Stewart *et al.*, 1977) contain chlorite in addition to allophane, gibbsite, and halloysite (Russell *et al.*, 1981).

In Japan, Andosols that contain large amounts of 2 : 1 layer silicates and the 2 : 1 to 2 : 1 : 1 layer silicate intergrades were found in Hokkaido, northern and western Honshu, and Kyushu (Wada, 1980). They contain those layer silicates in amounts up to 50% or more of their clay fraction, and some of them do not contain allophane and imogolite. The latter soils show a strong exchange acidity owing to the presence of exchangeable Al (Shoji and Ono, 1978; Saigusa *et al.*, 1980). Their occurrence is fairly extensive but is still localized in the respective districts. The amount of the layer silicates is usually higher in the A horizon than in the B horizon. This is expected from the concentration of the layer silicates that would occur in the A horizon if the Al-humus complex formation retards the formation of allophane and imogolite. The difference in the content of the layer silicates between the A and B horizons in a soil is, however, smaller than that between the soils derived from different volcanic ashes.

Soils containing a large amount of the layer silicates are derived from rhyolitic, dacitic, and andesitic ashes but not from basalt–andesitic and basaltic ashes (Shoji, 1983). They also contain substantial amounts of hornblende and quartz in the fine-sand fraction and/or quartz in the silt and clay fractions. This feature was interpreted as indicating that the formation of the 2 : 1 and 2 : 1 : 1 layer silicates occurs from dacitic ashes (Wada and Aomine, 1973). Inoue (1981) reviewed relevant studies and suggested the possibility that the fine-grained quartz is brought into Andosols as eolian dusts (wind-blown loess) from China and that the mica brought in at the same time can be a source material of the layer silicates. Mizota (1982, 1983) obtained the oxygen isotopic ratio of quartz in the two Andosols and the K–Ar age of K in the <20 μm fraction, which indicate their origin as eolian dusts. Whether all quartz that associates with hornblende and the layer silicates is accounted for by the admixture of eolian dusts awaits further study. On the other hand, Yamada and Shoji (1982) found a rather large variation of the K content (1.2 to 6.4% as $K_2O$) and its negative correlation with the Ca content in weathered volcanic glass fragments in the Andosols in which the layer silicates predominate. They inferred that the enrichment of volcanic glass with K leads to the formation of illite and subsequently to that of the 2:1 and 2:1:1 layer silicates. There was, however, no explanation for the quartz present in substantial amounts in all these samples.

## VI. Electric Charge Characteristics of Andosols

Andosols are noted as soils with variable electric charge. Figure 10 shows the development of negative and positive charges at different pH's and electrolyte concentrations for Andosols and weathered pumices. They are different in the major ion-exchange materials. The measurement of electric charge was carried out by determining the retention of $NH_4^+$ and $Cl^-$ by equilibrating the sample with $NH_4Cl$ solution at an appropriate concentration (0.1 to 0.005 $M$) and pH (4.5 to 8). The sample was saturated first with $NH_4^+$ and $Cl^-$ using 1 $M$ $NH_4Cl$, but the excess salt was not removed by washing. As illustrated in Figure 10, the "base"-holding capacity of Andosols varies very much with pH and electrolyte concentration of the solution. Conventional cation-exchange capacity CEC methods use, e.g., 1 $M$ $NH_4CH_3COO$ buffered at pH 7 for saturation of index cation, where the excess salt is removed by washing with water and/or alcohol. It is evident that the conventionally measured CEC of Andosols have no precise meaning and have little use in assessing the ability of the soils to retain "bases" in the field.

The magnitude of the effects of pH and electrolyte concentration on the negative charge of Andosols and weathered pumices depends on the kind of major ion-exchange materials (Figure 10). This is more clearly seen in the functional relationship derived from a linear regression analysis of the data:

$$\log \sigma^- = a\text{pH} + b \log C + c \qquad [1]$$

where $\sigma^-$ is the amount of negative charge (meq/100 g soil) and $C$ the $NH_4Cl$ concentration of the equilibrium solution ($M$). The coefficients $a$, $b$, and $c$ are constants for each soil, and their values for soils and weathered pumices, which are different in the major ion-exchange materials, are given in Table 9. These values can be used to estimate the magnitude of the effects of pH and electrolyte concentration on the amount of negative charge. It depends on the kind of ion-exchange material and decreases in the order: allophane and imogolite with a $SiO_2/Al_2O_3$ ratio close to 1.0, Al-humus complexes, 2:1 and 2:1:1 layer silicate intergrades > allophane with a $SiO_2/Al_2O_3$ ratio of about 2.0 > mica, chlorite, and vermiculite > kaolinite and vermiculite–chlorite intergrades > smectite.

The CEC value of imogolite, 20 to 30 meq/100 g, was obtained in $10^{-2}$ to $10^{-1}$ $M$ salt solution and at pH 7.0 (Wada, 1977; Theng et al., 1982). Almost all its negative charge is pH dependent (Figure 10) and probably arises from $H^+$ dissolication from the Si–OH groups that protrude inside of

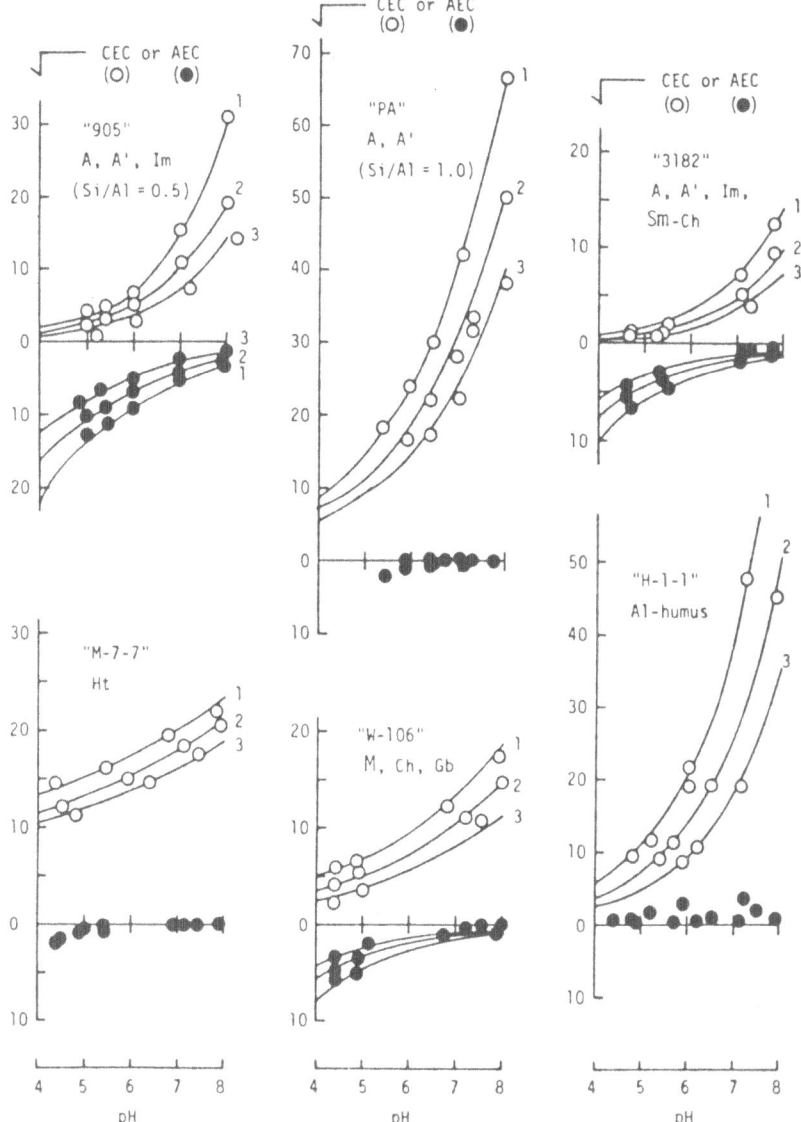

**Figure 10.** Electric charge characteristics of Andosols ("905," "3182," "W-106," "H-1-1") and weathered pumices ("PA" and "M-7-7"), which differ in major ion-exchange materials (see Table 8 for abbreviation), determined using $NH_4$ and $Cl$ as index ions and at different $NH_4Cl$ concentrations: $1 = 0.1$ $M$; $2 = 0.02$ $M$; and $3 = 0.005$ $M$. The amounts of $\sigma^-$ and $\sigma^+$ are expressed as meq/100 g soil. From Wada (1980), in *Soils with Variable Charge*, pp. 87–107. Copyright © 1980 by the New Zealand Society of Soil Science. Used with permission.

**Table 9.** Electric Charge Characteristics of Soils and Weathered Pumices Different in Major Ion-Exchange Materials[a]

| Sample | Major ion-exchange[b] materials | Negative charge[c] | | | | Positive charge[c] | | | |
|---|---|---|---|---|---|---|---|---|---|
| | | $a$ | $b$ | $c$ | $R^2$ | $a'$ | $b'$ | $c'$ | $R^2$ |
| Andosol B } | A, A', Im[d] | 0.31 | 0.25 | −0.72 | 0.99 | −0.21 | 0.20 | 2.36 | 0.99 |
| Andosol B } | | 0.34 | 0.27 | −0.97 | 0.98 | −0.25 | 0.25 | 2.30 | 0.94 |
| Weathered pumice | A, A'[e] | 0.22 | 0.17 | 0.22 | 0.99 | – | – | – | – |
| Andosol Bb } | A, A', Im, Sm–Ch or Vt–Ch | 0.33 | 0.23 | −1.26 | 0.99 | −0.23 | 0.18 | 2.08 | 0.99 |
| Andosol Bb } | | 0.25 | 0.27 | −0.29 | 0.99 | −0.23 | 0.23 | 2.46 | 0.97 |
| Andosol A₁ | Al(Fe)-humus | 0.29 | 0.27 | −0.17 | 1.00 | – | – | – | – |
| Andosol A₁b } | Al(Fe)-humus, A, A', Im | 0.34 | 0.31 | −0.31 | 0.99 | – | – | – | – |
| Andosol A₁b } | | 0.31 | 0.32 | −0.29 | 0.99 | – | – | – | – |
| Andosol Bb | M, Ch, Gb, Vt | 0.15 | 0.14 | 0.25 | 1.00 | −0.24 | 0.21 | 2.06 | 0.95 |
| Weathered pumice | Ht | 0.06 | 0.08 | 0.94 | 0.97 | – | – | – | – |
| Red-Yellow soil B | Kt, Vt–Ch | 0.09 | 0.08 | 0.47 | 0.96 | −0.25 | 0.14 | 1.81 | 0.98 |
| Red-Yellow soil B | Sm | 0.02 | 0.03 | 1.41 | 0.94 | – | – | – | – |

[a] Adapted from Wada and Okamura (1980) and Okamura and Wada (1983).

[b] Abbreviations: see footnote to Table 8 except for Kt = kaolinite.

[c] See equations [1] and [2] in text for $a$, $b$, and $c$ and $a'$, $b'$, and $c'$. $R^2$ is the determination coefficient for the regression.

[d] $SiO_2/Al_2O_3$ molar ratio = about 1.0.

[e] $SiO_2/Al_2O_3$ molar ratio = about 2.0.

the tube (Figure 8). The CEC values varied from 10 to 50 meq/100 g for allophanes with the $SiO_2/Al_2O_3$ ratio between about 1 and 2 (Wada, 1980; Gonzales-Batista et al., 1982; Theng et al., 1982). The difference in the negative charge characteristics between two allophanes with the $SiO_2$ /$Al_2O_3$ ratio of about 1 and 2 suggests the difference in the origin of negative charge (Figure 10; Table 9.) In both the allophanes, negative charge arises from the Si–OH group in the Si–O tetrahedra sheet that constitutes the wall of the spherule (Figure 6), but additional negative charge possibly arises from the substitution of Si with Al in the Si–O sheet in the allophane with the $SiO_2/Al_2O_3$ ratio of 2.

In Andosols containing allophane and imogolite, smectite or vermiculite, even if the latter is present in a substantial amount, does not exhibit constant negative charge (Figure 10). These Andosols and those containing only allophane and imogolite show very similar negative charge characteristics (Table 9). This suggests that the constant negative charges on smectite or vermiculite are balanced by polymer hydroxy-Al cations that carry variable positive charge.

Table 10 illustrates the effect of pH on the magnitude of negative charge on different humus complexes and humic acids. The pH effect is marked for Andosols containing Al-humus complexes and is less marked for an Andosol containing Al- and Ca-humus complexes and halloysite. A similar variation in the pH dependence is observed for the two Andosol surface soils in Chile. On the other hand, the effect of pH is much less for a Chernozemic soil containing Ca- and Mg-humus complexes and humic acids I and II, and III, which were extracted from Andosols using NaOH and $Na_4 P_2O_7$, respectively, and purified by treatment with a $H^+$-saturated cation-exchanger. In these samples, the ionization of most COOH groups occurs at pH 4, and the development of negative charge between pH 4 and 7 is small. The difference between the Andosols and the humic acids demonstrates that polymer hydroxy-Al and partly hydroxy-Fe cations that were removed by the extraction and purification modify the charge characteristics of humic acid. A similar effect resulting from the removal of Al was found for humus in a Podzol B horizon (Martin and Reeve, 1958).

Only Andosols containing allophane and imogolite with a $SiO_2/Al_2O_3$ ratio of about 1 exhibit positive charge comparable with negative charge at field pH values (Figure 10). The soils containing iron oxides, gibbsite, and 2:1 and 2:1:1 layer silicate intergrades, and the weathered pumice containing halloysite exhibit positive charge only at and below pH 5.0. Positive charge does not appear in soils containing Al-humus complexes unless the soils contain allophane and imogolite. This observation has an important bearing on the retention of non-specifically adsorbing anions such as $NO_3^-$ in $A_1$ or Ap horizons of Andosols.

Like negative charge, the effect of pH and electrolyte concentration on the magnitude of positive charge is represented by an equation:

**Table 10.** Development of Negative Charge at Different pH Values Relative to That at pH 7.0 on Soils and Humic Acids[a]

| Sample | Major ion-exchange[b] materials | pH | | | | |
|---|---|---|---|---|---|---|
| | | 8.0 | 7.0 | 6.0 | 5.0 | 4.0 |
| Andosol A₁ | Al(Fe)-humus | 1.95 | 1.00 | 0.51 | 0.26 | – |
| Andosol A₁b | Al(Fe)-humus, A, A', Im | 1.99 | 1.00 | 0.50 | 0.25 | – |
| Andosol A₁b | Al(Fe)-humus, Vt–Ch | 2.18 | 1.00 | 0.46 | 0.21 | – |
| Andosol A₁b | Al(Fe)-humus, A, A', Ht | 1.30 | 1.00 | 0.77 | 0.59 | – |
| Chernozem Ap | Ca-humus, Sm, Vt | – | 1.00 | 0.92 | 0.83 | 0.64 |
| Humic acid (Andosol) | | | | | | |
| I | | – | 1.00 | 0.91 | 0.85 | 0.68 |
| II | | – | 1.00 | 0.92 | 0.81 | 0.63 |
| III | | 1.04 | 1.00 | 0.94 | 0.84 | 0.73 |

[a]Adapted from Wada and Okamura (1980), *Journal of Soil Science* 31:307–314. Copyright © 1980 by the British Society of Soil Science. Used with permission.
[b]Abbreviations: see footnote to Table 8.

$$\log \sigma^+ = a'\text{pH} + b' \log C + c'. \qquad [2]$$

Unlike $a$ and $b$, the values of $a'$ and $b'$ are similar between the Andosols (Table 4.9). This suggests the similarities in the nature of anion-exchange sites and ionization reactions in allophane, imogolite, and gibbsite:

$$\begin{bmatrix} & & H_2O \\ & \diagup & \\ Al & & \\ & \diagdown & \\ & & OH \end{bmatrix}^0 + H^+ \rightleftharpoons \begin{bmatrix} & & H_2O \\ & \diagup & \\ Al & & \\ & \diagdown & \\ & & OH \end{bmatrix}^{+1} . \qquad [3]$$

Oxisols and some Ultisols and Alfisols develop variable positive charge, but the $a'$ values calculated from published data were smaller than those of Andosols (Okamura and Wada, 1983).

## VII. Cation Exchange in Andosols

### A. Base Status and Cation Selectivity

As illustrated in Table 1, the content of exchangeable "base" is generally low in Andosols developed in humid, temperate regions, such as Japan, New Zealand, and the northwestern U.S. This is a feature common to soils called Dystrandepts that have a "base" saturation of less than 50% in subsoils. Hydrandepts developed in the humid to perhumid tropic regions, e.g., Hawaii, show a similar feature. Though it is difficult to measure and define "base" saturation for soils with variable charge, the low "base" content of these soils is expected from the fact that their negative charge is smaller with decreasing pH and electrolyte concentration (Figure 10). In other words, their cation-exchange sites show a very high selectivity for $H^+$. Therefore, the soils show no strong acidity, despite their low to very low base saturation, as illustrated in Table 1. Exceptional Andosols are those containing large amounts of layer silicates, either halloysite or 2:1 layer silicates.

Even in the humid, temperate regions, Andosols at an early stage of development, e.g., the soils at site 220 in Table 1, show a relatively high "base" saturation. Soils called Eutrandepts developed in relatively arid regions also show a high "base" saturation. This is due to the reduced leaching of "bases" but also to the negative charge characteristics of halloysite (Figure 10) and possibly of related poorly crystalline minerals.

The selectivity with which the soil retains a cation on the exchange site over other cations is important to assess the fate of added cations. There is relatively little information on the cation selectivity of Andosols, particularly those with well-characterized ion-exchange materials. Hunsaker and Pratt (1971) reported a preference for Ca in the Ca-Mg exchange with

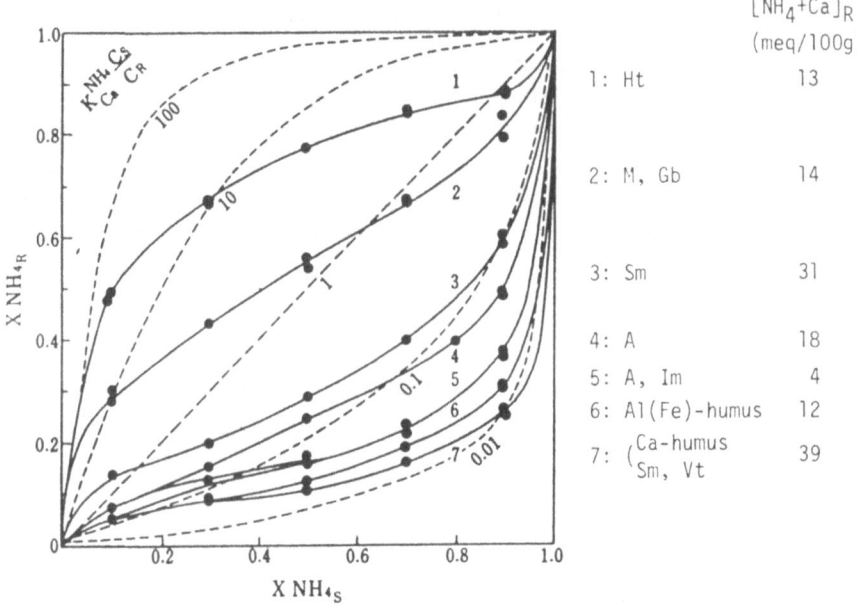

**Figure 11.** NH$_4$–Ca exchange equilibria for soils [2 = Andosol (B); 3 = Red-Yellow soil B; 5 = Andosol (B); 6 = Andosol A$_1$; 7 = Chernozem Ap] and weathered pumices (1 and 4) measured at C$_S$ = 0.01 $N$ and at pH = 4.7 to 6.3. The major ion-exchange materials (see Table 8 for abbreviation) and the NH$_4$ plus Ca retention (meq/100 g) for each sample are listed on the right-hand side of the diagram. From Wada (1981), in *Adsorption Phenomena in Soil—Fundamental and Application*, pp. 5–57. Copyright © 1981 by Hakuyusha Co. Ltd., Tokyo. Used with permission.

an allophane and a Dystrandept. Galindo and Bingham (1977) reported that K and Ca were preferentially adsorbed in the K–Na exchange and the Ca–Mg exchange with Chilean Dystrandepts, respectively. On the other hand, Wada and Abd-Elfattah (1978) found little difference in the selectivity between Ca and Mg with an Andosol B horizon sample.

Figure 11 shows the NH$_4$–Ca exchange for Andosol, weathered pumice, and Red-Yellow and Chernozem soil samples. In this determination the equivalent fractions of NH$_4$ ($X_{NH_4R}$) on exchange sites were measured by equilibrating the sample with 0.01 $N$ NH$_4$Cl · CaCl$_2$ containing NH$_4$ at different equivalent fractions ($X_{NH_4S}$). It is evident that when $X_{NH_4R} > X_{NH_4S}$ on the plot, the sample exhibits higher selectivity for NH$_4$ than Ca and vice versa. The broken-line curve in Figure 11 indicates the $X_{NH_4R}$–$X_{NH_4S}$ relation that would be expected when all cation-exchange sites in a soil have a specified value of NH$_4$/Ca selectivity. The latter selectivity is equated with K$_{Ca}^{NH_4}$($C_S/C_R$) by applying the law of mass action to the NH$_4$-Ca exchange, where the $K_{Ca}^{NH_4}$ is the equilibrium constant for the $NH_4$–Ca exchange and $C_R$ and $C_S$ the sum of equivalent concentrations of $NH_4$ and Ca in the solution and in the diffuse double-layer, respectively.

Figure 11 shows that each sample contains cation-exchange sites with different $NH_4$/Ca selectivities, and $NH_4$ is at first adsorbed on sites with higher $NH_4$/Ca selectivity and then on those with lower selectivity. A similar finding was made for Chilean Andosols in the K–Ca exchange by Schalscha *et al.* (1975). The differences apparent in the $NH_4$/Ca selectivity between the samples most probably depends on the nature of their major cation-exchange materials. The $NH_4$/Ca selectivity increases in the order: Chernozem Ap (Ca-humus, smectite, and vermiculite), Andosol $A_1$ (Al-humus), Andosol B (allophane and imogolite with $SiO_2/Al_2O_3$ ratio of about 1.0) < weathered pumice (allophane with $SiO_2/Al_2O_3$ ratio of about 2.0), Red-Yellow soil B (smectite) < Andosol B (illite, vermiculite, chlorite) < weathered pumice (halloysite).

In the K–Ca exchange for Chilean Andosols, Schalscha *et al.* (1975) found that an increase in pH produced an increase in negative charge that was largely balanced by the increase in the amount of Ca adsorbed. In the study on the $NH_4$–Ca exchange (Okamura and Wada, 1984), the effect of pH on the $NH_4$/Ca selectivity for each sample was interpreted in terms of the effect of $C_R$. As predicted from the $NH_4$/Ca selectivity being determined by $K_{Ca}^{NH_4}$ $(C_S/C_R)$, the measured $NH_4$/Ca selectivity increased with increasing $C_S$ and decreased with increasing $C_R$ and hence with increasing pH. The effect of the kind of cation-exchange material as illustrated in Figure 11 was, however, not accounted for by variation in $C_R$. Other factors such as the origin of negative charge, the steric feature around the exchange sites, clay–humus interaction, etc., may affect the cation selectivity.

Thus, Andosols with variable charge show a marked reduction in the retention of exchangeable "bases" with decreasing pH and electrolyte concentration, and this is more marked for K and $NH_4$ than for Ca. These two characteristics act together and result in weak retention of K and $NH_4$ under leaching conditions. The weak $NH_4$ retention would also be related to the low production capability of some Andosols used as paddy, where the surface soil is reduced by seasonal flooding and $NH_4$ is the main available form of nitrogen. Paddy rice growing on an allophanic soil also showed a high nitrogen absorption/dry matter production ratio at early stages of growth, as compared with other soils, which was related to a high $NH_4$ concentration in the soil solution (Seino *et al.*, 1976).

## B. Micronutrient Status and Specific Adsorption of Heavy Metals

In New Zealand, small additions of Co, Cu, and Se were found to be beneficial to livestock on soils from recent rhyolitic ashes (Taupo, Kaharoa, and Rotomahana ashes). These microelement additions were not needed in the Rangipo district, where Taupo ash had been dusted with small amounts of andesitic Ngauruhoe ash (Gibbs, 1968). Kobayashi (1981) pointed out that Co-deficiency in livestock occurs even on Andosols from pyroxene-andesitic ashes when they are young and coarse, because Co is chiefly present in unweathered primary minerals.

Micronutrient status in Andosols has not been fully elucidated in relation to their unique soil formation. The status of heavy metal cations in 67 Andosols in eastern Hokkaido was analyzed by successive dissolution using $H_2O_2$ and oxalate–oxalic acid without and with Mg metal (Miki et al., 1975). About 10 to 40% of Cu, 20 to 40% of Zn, and 20 to 40% of Co in these soils were dissolved. Their incorporation into humus complexes was indicated from a good correlation between the C content and the amounts of Zn, Cu, and Co dissolved by $H_2O_2$. There were also indications that Cu associated with "$SiO_2-Al_2O_3$" gel, and Zn and Co with free iron oxides. On the other hand, Shoji and Yamada (1977) noted the contrasting occurrence of Cu and Zn in relation to humus accumulation. Cu was concentrated in both present-day and buried surface horizons, whereas Zn was not. The cause of Cu deficiency in some Andosols was ascribed to the stable complex formation of Cu with humus (Tsutsumi et al., 1968), but also simply to the low Cu content (Mizota, 1981).

The adsorption of heavy metal cations by soils is important in determining their availability to plants and their movement through the soil. There are many studies on the adsorption of heavy metal cations with soils, soil minerals, and organic matter that have been reviewed together with recent findings for oxide minerals with variable charge (Parfitt, 1980). Andosols containing unique cation-exchange materials also show distinctive features in the adsorption of heavy metal cations.

Figure 12 illustrates several features of the adsorption of heavy metal cations on two soils; one is an Andosol (B) horizon containing allophane and imogolite, and the other is a Red-Yellow soil B horizon containing smectite as major cation-exchange materials. The soil samples were examined in their natural state or after saturated with Ca at about pH 7 (Ca saturated). In Figure 12 the $K_{Ca}^{Zn}$ is the selectivity coefficient defined by the equation

$$K_{Ca}^{Zn} = \frac{[Zn]_R[Ca]_S}{[Zn]_S[Ca]_R} \qquad [4]$$

for the Zn–Ca exchange, where $[Zn]_R$ and $[Ca]_R$ are the amounts of Zn and Ca adsorbed and $[Zn]_S$ and $[Ca]_S$ the concentrations of Zn and Ca in the equilibrium solution. The $K_{Ca}^{Zn}$ value indicated that the cation-exchange sites in the Andosol exhibit higher selectivities for Zn than those in the Red-Yellow soil, and this difference is more marked for the Ca-saturated samples than the natural samples. The Ca-saturated sample adsorbed larger amounts of Zn than the natural, weakly acid to acid sample of each soil at specified selectivities. This suggests a strong competition between Zn

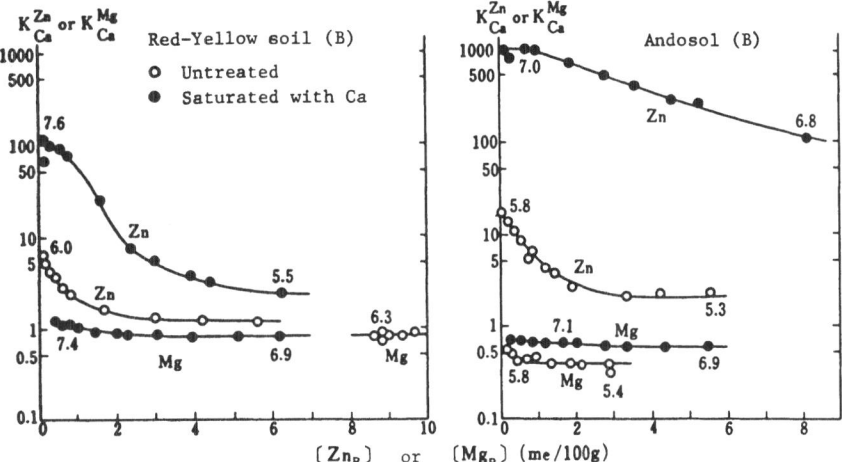

**Figure 12.** Zn–Ca and Mg–Ca exchange equilibria for Red-Yellow soil B ($[Zn+Ca]_R$ = 25.7 meq/100 g) and Andosol (B) ($[Zn+Ca]_R$ = 17.0 meq/100 g). The numbers along the curve indicate equilibrium pH. Adapted from Wada and Abd-Elfattah (1978), *Soil Science and Plant Nutrition* 24:417–426. Copyright © 1978 by the Japanese Society of Soil Science and Plant Nutrition. Used with permission.

and $H^+$ for potential cation-exchange sites. A good agreement between exchangeable Ca content and $[Zn + Ca]_R$ for each Ca-saturated sample indicated that the selective adsorption of Zn occurred on the cation-exchange sites and that the adsorption sites for "specifically" and "non-specifically" adsorbing cations are not different. The absence of specific adsorption for Mg is also illustrated in Figure 12.

Similar determinations for Andosols and other soils that are different in major ion-exchange materials with Cd, Co, Zn, Cu, and Pb led to the following observations (Abd-Elfattah and Wada, 1981):

1. Andosols containing allophane and imogolite are strong adsorbents and either Andosols or other soils that contain halloysite and/or are rich in iron oxides are the strongest adsorbents for most of the heavy metal cations. Andosols containing Al-humus complexes are also strong adsorbents for Pb and Cu.
2. The intrinsic affinities of heavy metal cations for Andosols as well as other soils increase in the order Cd, Co < Zn < Cu ≤ Pb. The highest $K_{Ca}^M$ value exceeds 10,000, where $M$ stands for a heavy metal cation.
3. The proportion of selective adsorption sites in the Ca-saturated samples (e.g., $K_{Ca}^M$ value >100) for Pb and Cu is about 50% or more in Andosols containing allophane and imogolite or halloysite and is less than 50% but larger than 20% in Andosols containing Al-humus

complexes. The corresponding figure for a Red-Yellow soil containing smectite is only 11 to 13%.

4. The Ca saturation of each soil sample increases both the amount of adsorbed heavy metal cation and the affinity with which the soil holds it.

These observations have important implications for the retention and availability of the heavy metal not only in Andosols but in other soils. The observed difference in the selective absorption of the heavy metal cations between the natural and Ca-saturated samples and those between soils containing different cation-exchange materials indicate that the cation-exchange sites developed at pH values higher than about 6.0 because of ionization of surface OH and COOH groups are important, but those resulting from isomorphous substitution in the 2:1 layer silicates are not. The adsorption of the heavy metal cation of high selectivity sites was formulated:

$$
\begin{array}{c}
- O^- \\
\phantom{} \\
- O^-
\end{array}
\quad Ca^{2+} + M^{2+} \rightleftharpoons
\begin{array}{c}
- O \\
\phantom{} \\
- O
\end{array}
\!\!\diagdown\!\!M + Ca^{2+}
\qquad [5]
$$

$$
\begin{array}{c}
- COO^- \\
\phantom{} \\
- COO^-
\end{array}
\quad Ca^{2+} + M^{2+} \rightleftharpoons
\begin{array}{c}
- COO \\
\phantom{} \\
- COO
\end{array}
\!\!\diagdown\!\!M + Ca^{2+}
\qquad [6]
$$

where $Ca^{2+}$ and $M^{2+}$ represent hydrated species and $M$ dehydrated species (Abd-Elfattah and Wada, 1981). The stability of the coordination complexes of heavy metals involving deprotonated OH or COOH groups as ligands would be greater than that of the ion-exchange complexes of hydrated heavy metal cations or Ca $^{2+}$. There were parallels between the selective absorption, hydrolysis ($p*K$ values), and precipitation as hydroxides ($pK_{so}$ values) of heavy metals.

## C. Adsorption of Organic Cations

Interaction of organic cations and soils has been gaining attention. Like interaction with inorganic cations, it arises mainly from an electrostatic interaction, but it is also affected by the size and shape of organic cations. Adsorption of alkylammonium cations that decreases with increasing molecular size on allophane was reported by Birrell (1961) and Theng (1972). They interpreted it in terms of physical adsorption, where the molecular size of the cation affects their packing in some interstitial pores. An "exclusion" of large organic cations from exchange sites of allophane was also reported for cetylpyridinium ion (Greenland and Quirk, 1962), o-phenanthroline (Aomine and Otsuka, 1968), and paraquat ions (Knight and Tomlinson, 1967). Studies on adsorption of various organic cations by smectite and vermiculite showed that large organic cations are often more strongly adsorbed than inorganic cations, depending on their molecular configuration, which was attributed to the increased contribution of van der Waals forces (Theng *et al.*, 1967).

Table 11 shows one example of exchange equilibria between Ca and $NH_4$, alkylammonium ion, or piperidinium ion on Andosol and other soil and clay samples, which are different in ion-exchange materials. In this table, $X_{AR}$ represents the equivalent fraction of adsorbed $NH_4$ or the organic cations (A), when each Ca-saturated sample was washed repeatedly with a 1:1 mixture of 0.05 $M$ $CaCl_2$ and 0.1 $M$ $NH_4Cl$, alkylammonium chloride, or piperidinium chloride. Because the equivalent fraction of A in the equilibrium solution $X_{AS}$ is 0.5, the $X_{AR}$ value gives a measure of the A/Ca selectivity.

The data in Table 11 show that each sample exhibits different A/Ca selectivity depending on its ion-exchange materials. Bentonite and a Red-Yellow soil containing smectite and a halloysite clay exhibit high selectivities for the organic cations. The bentonite has particularly high selectivities for alkylammonium ions that are higher than $NH_4$ and increase with increasing molecular size of cations. The Andosols containing allophane and imogolite exhibit low $NH_4$/Ca selectivities (Figure 11). They exhibit further lower selectivities for larger organic cations; the largest alkylammonium $(CH_3)_4N$ and piperidinium ions are not retained on their cation-exchange sites. The constancy of the total amount of adsorbed Ca and A, $[Ca+A]_R$, indicates that the adsorption of organic cations occurred on the cation-exchange sites. It is considered that the large organic cations do not have access or only have difficult access to all exchange sites in allophane and imogolite and are "excluded." Such a "molecular sieve" effect likely occurs in allophane and imogolite that consist of very minute hollow spherules (Figure 5) and tubes (Figure 7), respectively. The two Andosols containing humus also show very low selectivities for the organic cations, but unlike those containing allophane and imogolite, they exhibit selectivity for the alkylammonium ion that does not depend much on its molecular size and higher selectivity for the piperidinium ion. Their low A/Ca selectivities may be attributed to strong affinities of $Ca^{2+}$ with ionized carboxyl groups.

These observations have implications to the fate of added organic chemicals in Andosols that behave as cations in the soil solution. Data show a close correlation between the reactivity of chemicals and the clay mineralogy of the soils. The phytotoxicity of paraquat* (Knight and Tomlinson, 1967) and cartup* (Inoue *et al.*, 1980) is high in soils containing allophane and imogolite, intermediate in soils containing kaolinite, halloysite, and/or illite, and low to very low in soils containing smectite.

---

*Paraquat (1:1' = dimethyl 4:4' = dipyridylium dichloride) and cartup (1,3-Bis (carbamoylithio)-2-(*N,N*-dimethyl)aminopropane) are trade names.

**Table 11.** A Summary of Data on Ca–A (A; NH₄, Alkylammonium, or Piperidinium Ion) Exchange on Soils and Clays That Differ in Ion-Exchange Materials[a]

| Sample | Major ion[b] exchange materials | $[Ca + Al]_R$[c] (meq/100g) | $X_{AR}$[c] | | | | | |
|---|---|---|---|---|---|---|---|---|
| | | | NH₄ | MMA[d] | DMA[d] | TrMA[d] | TeMA[d] | Pip[d] |
| Bentonite | Sm | 64–65.5 | 0.31 | 0.38 | 0.55 | 0.75 | 0.85 | 0.83 |
| Red-Yellow soil B | Sm | 31.5–34 | 0.45 | 0.37 | 0.43 | 0.36 | 0.44 | 0.45 |
| Weathered andesite | Ht | 12–13 | 0.58 | 0.52 | 0.32 | 0.28 | 0.37 | 0.23 |
| Andosol B | A, A', Im[e] | 10 | 0.26 | n.d.[f] | n.d. | n.d. | 0 | 0 |
| Andosol B | A, A', Im[e] | 17 | 0.20 | 0.10 | 0.05 | 0.02 | 0 | 0 |
| Weathered pumice | A, A'[g] | 33–34 | 0.38 | n.d. | n.d. | n.d. | 0.11 | 0.06 |
| Andosol A₁ | Al-humus | 33–36 | 0.08 | 0.03 | 0.03 | 0.02 | 0.04 | 0.06 |
| Chernozem Ap | Ca-humus Sm, Vt | 38–39 | 0.18 | n.d. | 0.11 | n.d. | 0.14 | 0.19 |

[a] Adapted from Wada and Tange (1984).
[b] Abbreviations: see footnote to Table 8.
[c] $[Ca+Al]_R$ and $X_{AR}$: See text.
[d] MMA = monomethylammonium; DMA = dimethylammonium; TrMA = trimethylammonium, TeMA = tetramethylammonium, Pip = piperidinium.
[e] $SiO_2/Al_2O_3$ molar ratio = about 1.
[f] n.d. = no data.
[g] $SiO_2/Al_2O_3$ molar ratio = about 2.

The "exclusion" of a large organic cation can be used for the tests of allophane and imogolite by combining its metachromasis. Toluidine blue $(CH_3)_2N^+C_6H_3NSC_6H_2(CH_3)NH_2$ shows the metachromasis, a color change from blue to purplish red when adsorbed on negatively charged surfaces. Nomoto *et al.* (1955) attempted to use this metachromasis for determination of the sign of electric charge on the soil and found that the metachromasis occurred for most soils derived from granite, andesite, and sedimentary rocks, whereas it occurred for 20 of 44 soils derived from volcanic ash. Tests for Chilean and Ecuadoran soil samples derived from volcanic ash that contained or lacked allophane and imogolite showed that the absence of the metachromasis can be used as a test for the dominance of allophane and imogolite (Wada and Kakuto, 1985).

## VIII. Reaction with Phosphates

As shown in Table 1, Andosols adsorb increasing amounts of phosphate as the process of soil formation proceeds. The phosphate sorption coefficient in this table is determined by adding 20 ml neutral 2.5% ammonium phosphate solution to 10 g of air-dry soil; the value of 1500 has been used in Japan to determine whether or not soils are derived from volcanic ash. The phosphate sorption coefficient of Andosols, except coarse-textured ones, is usually 1500 to 2500 (Table 4.1). The value of 1500 has been found to correspond to 90% of Blakemore's phosphate retention (Smith[*]) in tests of many Japanese Andosols (Amano, 1983).

Figure 13 illustrates that humus and allophane in Andosols relate to phosphate adsorption, irrespective of their use for upland crops and paddy rice. Allophane and imogolite have been considered as major materials contributing to the phosphate sorption in Andosols. Rajan (1975), Parfitt and Henmi (1980), and Theng *et al.* (1982) found that phosphate adsorption curves leveled off at 200 to 600 μmol $Pg^{-1}$ for imogolite and 5 to 10 μmol $Pg^{-1}$ for halloysite at P concentrations of 1 to $1.5 \times 10^{-4}$ $M$ in the presence of $10^{-2}$ $M$ KCl or $CaCl_2$. The concomitant release of sulfate, hydroxyl, and silicate suggests that phosphate is adsorbed by more than one mechanism, i.e., by exchanging $SO_4^{2-}$ and $OH^-$ bonded with Al and adsorbed silicate and by displacing $SiO_4$ groups of aluminosilicates, depending on whether the P concentration in solution is low (up to $10^{-2}$ $M$) or high, respectively (Rajan, 1975).

The importance of humus, more exactly Al- and Fe-humus complexes, and allophanelike constituents in phosphate adsorption was indicated by the following observations, as reviewed by Wada (1980): (1) the reduction of P adsorption by Andosols after treatments with $H_2O_2$ and/or dithionite–

---

[*]See footnote on p. 174.

**Table 12.** Phosphate Adsorption Data for Andosols[a]

| Soil group (sample number) | Phosphate adsorbing material[c] | Phosphate adsorption[b] | | |
|:---:|:---:|:---:|:---:|:---:|
| | | α | β | γ |
| I (8) | Al-humus>Fe-humus>A, (Im)>Fe-ox | −0.03 to 0.01 | 0.40 to 0.48 | −0.33 to 0.47 |
| II (7) | A, Im>Al-humus, A′, Fe-ox | −0.07 to −0.02 | 0.39 to 0.47 | 0.41 to 0.66 |
| III (3) | A, Im>A′, Fe-ox | −0.13 to −0.08 | 0.28 to 0.40 | 0.72 to 0.96 |

[a]Adapted from Gunjigake and Wada (1981).
[b]See equation [7] in text for α, β, and γ.
[c]Abbreviations: see footnote to Table 8.

citrate, (2) a high positive correlation between the humus content and the P adsorption capacity (Figure 13), and (3) a large P adsorption by Al horizons of Andosols in which allophane and imogolite are absent, or nearly absent.

Table 12 shows a summary of phosphate adsorption data for Andosols, which are different in the composition of phosphate-adsorbing materials at different pH values (4 to 8.5) and P concentrations ($10^{-3}$ to 2 $M$). A linear regression analysis of the data showed that the relationship between the amount of retained P ($Y$; mmol $g^{-1}$) and the equilibrium pH and P concentration ($C$; $M$) is represented by an equation

$$\log Y = \alpha \, pH + \beta \log C + \gamma, \qquad [7]$$

where α, β, and γ are coefficients constant for each soil, the values of which are given in Table 12. The coefficient of determination $R^2$ indicated that the 97 to 100% variation in retained P can be explained by pH and P concentration of the solution.

As described before, the similar functional relationship holds for the retention of Cl⁻ on the two soils of group III in Table 12. The $a$ (−0.25, −0.21) and $b$ (0.25, 0.20) values found for the Cl⁻ retention by the two soils (Table 4.9) are, however, different from the α (−0.11, −0.13) and β (0.40, 0.34) values given for phosphates, indicating the differences in the non-specific and specific adsorption of anions. The α and β values in Table 12 denote the differences in the dependence of P retention on the pH and P concentration between the different groups of soils. Those containing

**Figure 13.** Relation between the C content and phosphate adsorption of Andosols used for paddy rice (circle) and upland crops (square), which differ in containing (open) and not containing (closed) allophane and imogolite. From Mizota *et al.* (1982), *Geoderma* 27:225–237. Copyright © 1982 by Elsevier Science Publishers, Amsterdam. Used with permission.

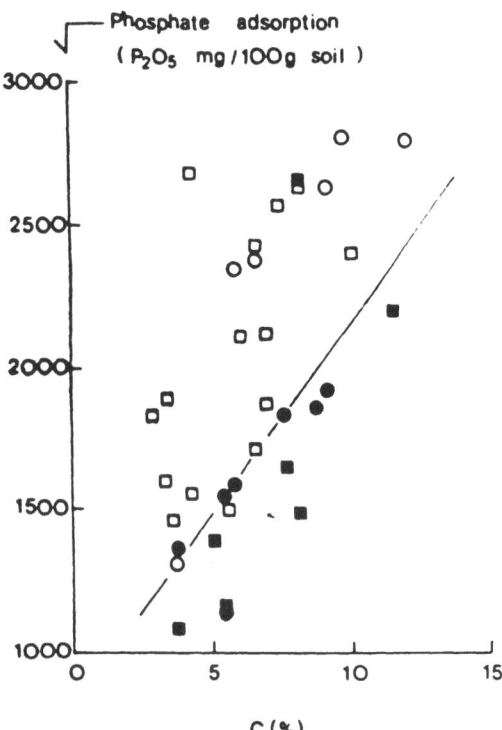

allophane and imogolite show appreciable pH dependence, whereas those containing Al- and Fe-humus complexes do not. On the other hand, the latter soils show greater P concentration dependence than the former soils. The molar reactivity of different forms of Al and Fe in Andosols toward phosphate (P/Al or Fe) are different pH's and phosphate concentrations was also estimated from the adsorption data by a multiple linear regression analysis (Gunjigake and Wada, 1981).

The differences in the reaction with phosphate between the different forms of active Al and Fe gives a clue to the availability of adsorbed phosphate and the reaction mechanism. The ligand exchange between $H_2O$ bonded to Al in Al-humus complexes and phosphate is shown as reactions [1] and [2] in Table 13, where $(^{+0.5}H_2O)_2$-Al represents the positively charged portion of hydroxy-Al polymers. These reactions explain simultaneous adsorption of counter cations, coupled with adsorption of phosphate by Andosols (e.g., Schalscha *et al.*, 1972, 1974; Ono and Uchida, 1979), the equal reactivity toward $H_2PO_4^-$ and $HPO_4^{2-}$, and the high P concentration dependence, as observed with the soils containing Al-humus complexes. The pH-dependent reactivity of Al in allophane and imogolite suggests a reaction that occurs on their positive charge sites (reaction [3] in Table 13). Such positive charge sites are present on

**Table 13.** A Schematic Presentation of Reactions of Active Al and Fe with Phosphate[a]

| Reaction | Scheme of reaction |
| --- | --- |
| [1] | $COO^- \overset{+0.5H_2O}{\diagup} Al + Na^+ + H_2PO_4^- \rightarrow COO^{-\,+}Na^{-0.5}H_2PO_4 \overset{}{\diagup} Al + H_2O$ <br> $\underset{+0.5H_2O}{\diagdown} \qquad\qquad\qquad\qquad\qquad\qquad \underset{+0.5H_2O}{\diagdown}$ |
| [2] | $COO^- \overset{+0.5H_2O}{\diagup} Al + 2\,Na^+ + HPO_4^{2-} \rightarrow COO^{-\,+}Na\;Na^{+\,-0.5}H_2PO_4 \overset{}{\diagup} Al + H_2O$ <br> $\underset{+0.5H_2O}{\diagdown} \qquad\qquad\qquad\qquad\qquad\qquad\qquad \underset{-0.5OH}{\diagdown}$ |
| [3] | $OH^- \overset{+0.5H_2O}{\diagup} Al + Na^+ + H_2PO_4^- \rightarrow \overset{-0.5}{H_2PO_4} \diagup Al + Na^+ + OH^- + H_2O$ <br> $\underset{+0.5H_2O}{\diagdown} \qquad\qquad\qquad\qquad\qquad \underset{+0.5H_2O}{\diagdown}$ |
| [4] | $\overset{+0.5H_2O}{\diagup} Al + Na^+ + H_2PO_4^- \rightarrow Na^+ \overset{-0.5}{H_2PO_4} \diagup Al + H_2O$ <br> $\underset{-0.5OH}{\diagdown} \qquad\qquad\qquad\qquad\qquad\qquad \underset{-0.5OH}{\diagdown}$ |

[a] Adapted from Gunjigake and Wada (1981), *Soil Science* 132:347–352. Copyright © 1981 by The Williams & Wilkins Co., Baltimore. Used with permission.

allophane and imogolite, but not on Al- and Fe-humus complexes. A ligand exchange reaction may occur for $H_2PO_4^-$ and $H_2O$ bonded to Al in allophane and imogolite (reaction [4] in Table 13), but not for $HPO_4^{2-}$, in view of the marked and negative pH dependence of the reaction.

## IX. Physical and Mechanical Properties

As illustrated in Table 1, Andosols containing large amounts of humus and clay have a low bulk density, i.e., 0.65 to 0.25 (sites 217 and 206), whereas those dominated by coarse, vitric materials that represent an early stage of soil formation have a bulk density higher than 0.8 (site 220). Gradwell (1974, 1976) and Maeda *et al.* (1983) found that the boundary bulk density value that determines whether or not soils are derived from volcanic ash in New Zealand and Japan is 0.85 g cm$^{-3}$. Bulk density of less than 0.85 g cm$^{-3}$ has been adopted as one of the diagnostic properties of soils in which amorphous material dominated in the exchange complex (Soil Survey Staff, 1975).

The low bulk density of some Andosols is partly due to the high content of humus and its low particle density, i.e., 1.4 to 1.8 g cm$^{-3}$. The low particle density of volcanic glass, 2.4 g cm$^{-3}$ (Wada and Wada, 1977), also contributes to the low bulk density of Andosols. However, the low bulk density of Andosols mainly reflects their high porosity. This is well illustrated in Figure 14, which shows a comparison of three-phase composition of surface horizons of Andosols and other soils (Misono, 1964). Andosols are small to very small in the volume of solids and large in the volume of pores in which the proportions of water and air vary in wide ranges. The solids volume is so small that the supply and retention of nutrients for plants can be a problem, whereas the large pore volume ensures aeration and water permeability and retention. The volume of solids varies little with the depth, whereas the volume of air and water decreases and increases with the depth, respectively.

The large pore space of Andosols containing allophane and imogolite is partly accounted for by the pore space (25 to 45%) in their structure units. The large pore space is, however, found not only in Andosols containing allophane and imogolite but also in those not containing these minerals but containing considerable amounts of humus. It is inferred that the micro pores are present as intra-unit pores in allophane, imogolite, and humus, and the medium to micro pores are created by interactions between them and other clay-, silt-, and sand-sized minerals, resulting in formation of stable aggregates. In this aggregate formation, polymer hydroxy-Al ions and allophane, allophanelike constituents, and imogolite, which carry positive charge and contain active Al, play an important role.

Andosols are normally permeable, and a high amount of water percolates through the soil in the humid to perhumid region. The profiles do

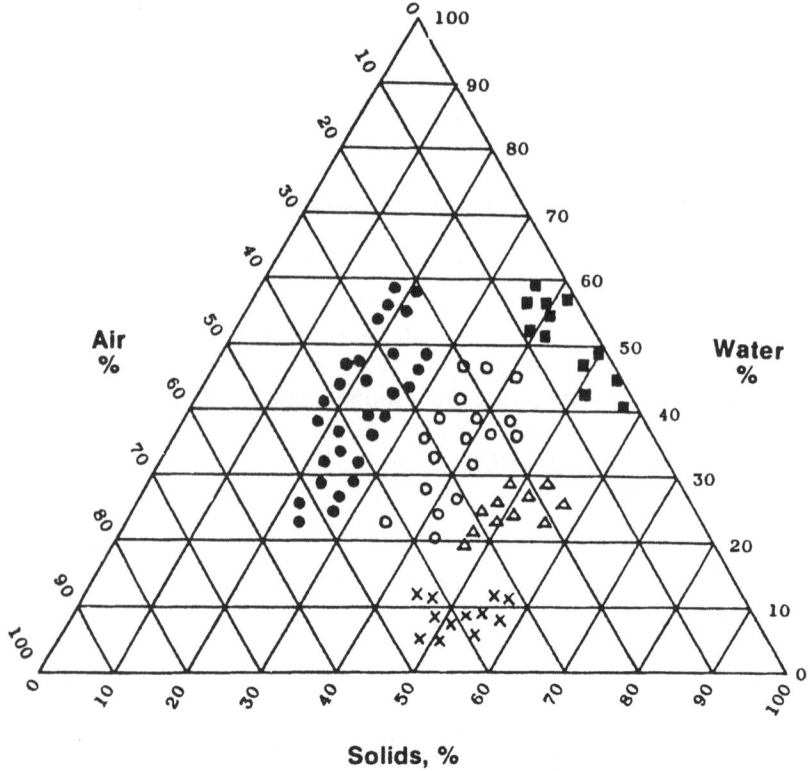

**Figure 14.** Three-phase diagrams for Andosols (closed circle), alluvial soils (open circle), paddy soils derived from alluvium (closed square), Red and Yellow soils (open triangle), and soils derived from dune sand (cross). From Misono (1964), in *Volcanic Ash Soils in Japan*, pp. 75–79. Copyright © 1964 by the Ministry of Agriculture and Forestry, Japanese Government. Used with permission.

not show, however, evidence of clay movement, indicating the formation of stable aggregates. The latter formation of stable aggregates is enhanced by drying that causes a difficulty for particle-size distribution analysis. Kubota (1976) demonstrated this by particle-size analyses of moist, air-dried, and oven-dried Andosols B and Red-Yellow soil B horizon samples and inferred that allophane contributes to the formation of stable aggregates. The degree of aggregation in the moist Andosol samples differed depending on the moisture condition under which they formed. Aggregates of allophane, allophanelike constituents, or imogolite and humus are, however, disrupted by $H_2O_2$ treatment and sonic oscillation. After $H_2O_2$ treatment, pH adjustment and sonic vibration are important for deflocculation and dispersion of clay. An alkaline medium (pH 10) can be used for soils containing allophane with a high $SiO_2/Al_2O_3$ ratio of 1.7 to 2 and/or halloysite or only layer silicates, whereas an acid medium (pH 4) is required for soils containing allophane and imogolite with the lower $SiO_2/Al_2O_3$ ratios with or without layer silicates, particularly 2:1 to 2:1:1 layer silicate intergrades. The clay content determined by repeating dispersion and sedimentation was in fairly good agreement with that determined by the method that includes dissolution of allophane and imogolite (Kuroboku Soken, 1983), though the clay content thus determined gave higher values than that estimated from the soil texture in the field.

Maeda *et al.* (1977) reviewed the data on plastic and liquid limits of Andosols and stated: (1) wet allophane soils have a high liquid limit, but also a high plastic limit, and hence a low plasticity index; (2) as the samples are gradually dried, the liquid limit decreases more rapidly than the plastic limit; and (3) highly allophanic soils become non-plastic before they reach the air-dry water content. They suggested that these features can be used to assess the intensity of allophanic characteristics of soils. Maeda *et al.* (1983) reached, however, a different conclusion, that plasticity does not uniquely characterize Andosols, by examining more recent data on Japanese Andosols whose clay mineral composition and humus contents were analyzed (Kuroboku Soken, 1983). The data are shown on a Cassagrande plot in Figure 15. The plot indicates that these Andosols do not have a particularly low plasticity index as reported by Wesley (1973) for Andosols in Java and by Maeda *et al.* (1983) for Andosols in Dominica. Exceptions are a few present-day surface soils, which have a lower plasticity index, possibly reflecting the effect of drying. The plot also indicates that soils containing allophane, imogolite, and halloysite and/or a large amount of humus have a high liquid limit value.

A high natural water content on 1/3 bar water is a feature common to Andosols. Available water content is usually defined as 1/3 bar water content minus 15 bar water content. Figure 16 shows the relation between 1/3 bar water content and available water content for Andosols in different regions. The relation is similar for Andosols in Alaska, Washington, Oregon, Chile, Ecuador, and dry parts of Hawaii. About 20 to 70% of 1/3 bar water in these soils is available, though the 1/3 bar water content varies in a wide range in Alaska and Chile and is low in Washington, Oregon, Ecuador, and dry parts of Hawaii. Andosols in perhumid parts of Hawaii, Hydrandepts, are high in 1/3 bar water, but the proportion of available water in the 1/3 bar water is intermediate (15 to 30%). Andosols in Japan are peculiar in that the available water content correlates little with the 1/3 bar water content, and only a small portion of the latter water ($\leq$15 to 20%) is available. This feature was also noted by Maeda *et al.* (1983) in comparison with Andosols in New Zealand.

The unique feature of the Hydrandepts of Hawaii (Figure 16) may be interpreted as indicating the effect of predominance of non-crystalline Al and Fe hydroxides rather than allophane and imogolite, whereas the difference between the Chilean and Japanese Andosols is difficult to explain on the basis either of the clay mineral composition and humus content or of

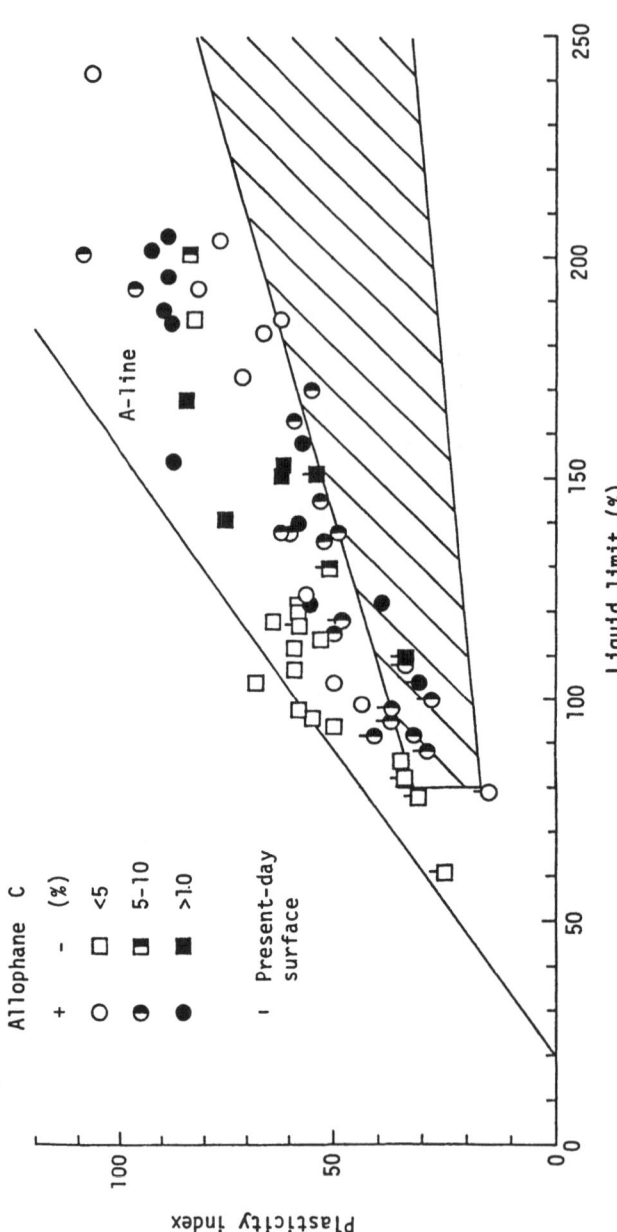

**Figure 15.** Cassagrande plasticity-chart for Japanese Andosols that differ in humus content and clay mineral composition. Based on data from Kuroboku Soken (1983). The plots for Andosols in Java (Wesley, 1973) and those in Dominica (Maeda *et al.*, 1983) fall in the area enclosed with oblique lines.

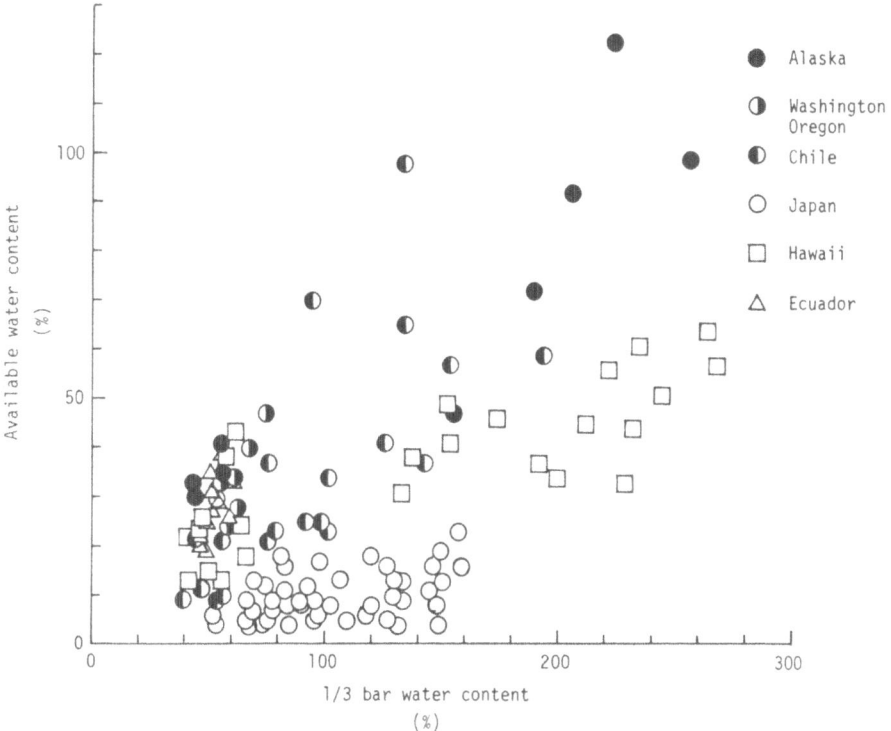

**Figure 16.** Relation between the contents of available water (1/3 bar water minus 1/15 bar water) and 1/3 bar water in Andosols of different regions. Based on data from Flach (1969), Sixth International Soil Classification Workshop (1984), Soil Survey Staff (1975), and Kuroboku Soken (1983).

the climates under which they were formed. The low available water content of the Japanese Andosols resulted from the higher 15 bar water content. As reviewed by Egawa (1977), Andosols in Japan had been considered to have high available water contents. Maeda and Soma (1979) showed, however, that some of these high values are associated with the method of measuring water retention. They found that the centrifuge method gives a lower water content at 15 bar than the pressure plate method, arising from compaction of a sample by centrifugal force. The available water content estimated by the centrifuge method amounted to twice or more that estimated by the pressure plate method. All the data shown in Figure 16 were, however, obtained by the pressure plate method. The cause or the mechanism by which a large amount of water is retained at 15 bar in Japanese Andosols awaits further study.

Andosols have distinctive engineering properties associated with their strong aggregate formation. Maeda *et al.* (1977) reviewed studies on those properties of allophane soils, and the following subsections are abstracted mostly from their review, unless otherwise stated.

## A. Compaction

Unlike soils containing crystalline clay minerals, undried allophane soils do not show a distinct maximum in the bulk density on compaction, which can be observed only by gradually drying the samples before measurement. The compaction curve obtained by repeating drying and wetting a sample forms hysteresis loops with only the wetting portion showing a well-defined density maximum. The latter maximum is noted when the soil was dried over 1 bar suction, and its value increases with the drying over 15 bar suction, which is the highest suction the subsoil would be subjected to under natural soil conditions with drying only by plant roots.

## B. Strength

Andosols can be stable in the undisturbed state and are relatively resistant to erosion. The strength of allophane soils disturbed by excavation and embarkment is remarkably lower than that of the undisturbed soils. Kitajima (1983) measured the amount of soil loss by erosion at the northern somma of Mt. Aso. It was 3.0 to 7.4 $m^3$ $ha^{-1}$ $yr^{-1}$ for pasture with natural grass and 2.4 to 3.9 $m^3$ $ha^{-1}$ $yr^{-1}$ for improved pasture. During the land preparation and cultivation for the improvement, the amount of soil loss increased to 9.5 to 15.5 $m^3$ $ha^{-1}$ $yr^{-1}$.

Allophane soils have the high bearing capacity to support buildings in spite of their high natural water content, and their unconfined compressive strength is in the range of 1.0 to 2.3 kg $cm^{-2}$. This is more than five times the strength of alluvial soils at the same water content. The strength of Andosols does not increase with depth, i.e., with increasing overburden pressure. Allophane soils have a moderate sensitivity; the measured ratio of undisturbed to remolded shear strength is 1:12. The lessening of strength by remolding is due to changes in water-holding characteristics and structure peculiar to allophane soils. On disturbance, water that was held firmly in the pores is released and free to flow. The soils least resistant to this softening are buried soils with high organic matter, especially those developed from old volcanic ashes. Subsoils containing exclusively spherical halloysite often show a very high sensitivity. A value as high as 140 or more was reported by Kita et al. (1969) and Smalley et al. (1980).

## C. Consolidation

Allophane soils are fairly compressible, once the preconsolidation pressure has been exceeded. It is easy to find the preconsolidation pressure for the undisturbed samples, and the values range from 1 to 3 kg $cm^{-2}$ and are not correlated with depth. The preconsolidation pressure often exceeds the overburden pressure owing to the strength of the aggregates.

# X. Conclusions

Volcanic ash and pumice as parent materials have two distinctive features: First, they are fine-grained or porous and have large specific surface area that enhances their interaction with environment and accelerates their weathering. Second, they contain large amounts of non-crystalline volcanic glass and other weatherable minerals. A rapid weathering of volcanic ash and pumice, particularly under humid conditions and at well-drained sites, results in loss of Si and "bases" by leaching and in accumulation of Al and Fe in weathering products. Different classes of non-crystalline and para-crystalline clay minerals, such as allophane, allophanelike constituents, opaline silica, and imogolite and Al- and Fe-humus complexes, form as products of soil-forming processes. Knowledge about the structure, composition, and morphology of allophane and imogolite and recognition of the nature of the humus complexes has provided a basis for interpreting their effect on the distinctive chemical and physical properties of Andosols. A better understanding of the soil-forming processes has been obtained by relating the formation and transformation of the non- and para-crystalline clay minerals to humus accumulation.

The exchange complex in Andosols was considered to consist dominantly of X-ray amorphous compounds of Al, Si, and humus. A proposition that it would be better defined in terms of the dominant presence of active Al rather than of amorphous materials has emerged (Wada, 1980; Leamy, 1983). The active Al forms humus complexes and constitutes a part of non- or para-crystalline aluminosilicates, such as allophane, allophanelike constituents, and imogolite, and possibly also a part of interlayer in 2:1 and 2:1:1 interlayer silicate intergrades. The active Al is a determinant not only of positive but of some negative variable charge characteristics of Andosols and reacts with phosphate and fluoride. It contributes to the accumulation or the preservation of humus and to the formation of stable aggregates. Fe as well as Al in humus complexes are active, but Fe in iron oxides is not, though Al and Fe in non-crystalline hydroxides are highly reactive in Andosols formed in the perhumid tropics.

Andosols may have formed under subhumid to arid climates or climates with pronounced dry seasons. Recent reports on soils in Kenya, Sudan, Rwanda, and Tanzania (e.g., Wielemaker and Wakatsuki, 1984) have, however, indicated that soil formation from volcanic ash, and hence the nature and properties of the resulting soil, in such climates are markedly different from those of Andosols in the humid temperate or the perhumid tropics. Though the relation with climate is not clear, some Ecuadoran soils may also fit in this category. They contain poorly crystalline clay minerals, not as yet well characterized, which give no indication of reactive Al. These soils and soils called Eutrandepts await further study on their genesis and characterization of clay minerals and humus complexes.

# References

Abd-Elfattah, A., and K. Wada. 1981. Adsorption of lead, copper, zinc, cobalt, and cadmium by soils that differ in cation-exchange materials. *J. Soil Sci.* 32:271–283.

Adachi, T. 1973. Studies on the humus of volcanic ash soils—regional differences of humus composition in volcanic ash soils in Japan. *Bull. Nat. Inst. Agr. Sci. Japan* B24:127–264. (In Japanese)

Amano, Y. 1983. Andisol proposal and Japanese volcanic ash soils—criteria for chemical properties. In: N. Yoshinaga (ed.), *Volcanic ash soils—genesis, properties, classification.* pp. 187–204. Hakuyusha, Tokyo, 204 pp. (In Japanese)

Aomine, S. 1972. Nitrogen fertility and humic matter of Chilean Andosols. *Soil Sci. Plant Nutr.* 18:105–113.

Aomine, S., and H. Otsuka. 1968. Surface of soil allophanic clays. *Trans. 9th Intl. Congr. Soil Sci.* 1:731–737.

Arai, S. 1983. Humus in volcanic ash soil. In: N. Yoshinaga (ed.), *Volcanic ash soil—Genesis, Properties, Classification.* pp. 73–98. Hakuyusha, Tokyo. 204 pp. (In Japanese)

Araragi, M., ande S. Ishizawa. 1972. Actinomycete flora of Japanese soils. *Bull. Nat. Inst. Agr. Sci. Japan* B23:147–255. (In Japanese)

Bascomb, C.L. 1968. Distribution of pyrophosphate-extractable iron and organic carbon in soils of various groups. *J. Soil Sci.* 19:251–268.

Birrell, K.S. 1961. The adsorption of cations from solutions by allophane in relation to their effective size. *J. Soil Sci.* 12:307–316.

Brydon, J.E., and S. Shimoda. 1972. Allophane and other amorphous constituents in a podzol from Nova Scotia. *Can. J. Soil Sci.* 52:465–475.

Claridge, G.G.C. 1981. Mineralogy of the soils of the Kingdom of Tonga. *Soils with Variable Charge, Programme and Abstracts.* pp. 176–177.

Colmet-Daage, F. 1969. Nature of the clay of some volcanic ash soils of the Antilles, Ecuador and Nicaragua. In: *Panel on Volcanic Ash Soils in Latin America.* Turrialba, Costa Rica. B.2.1–B.2.11.

Cradwick, P.D.G., V.C. Farmer, J.D. Russell, C.R. Masson, K. Wada, and N. Yoshinaga. 1972. Imogolite, a hydrated aluminum silicate of tubular structure. *Nature Phys. Sci.* 240:187–189.

Dudas, M.J., and M.E. Harward, 1975. Inherited and detrital 2:1 type phyllosilicates in soils developed from Mazama ash. *Soil Sci. Soc. Am. Proc.* 39:571–577.

Egawa, T. 1977. Properties of soils derived from volcanic ash. In: Y. Ishizuka and C.A. Black (eds.), *Soil derived from volcanic ash in Japan.* pp. 10–63. International Maize and Wheat Improvement Center, Mexico. 102 pp.

FAO/UNESCO. 1974. *Soil map of the world. Vol. I. Legend.* UNESCO, Paris. 59 pp.

Farmer, V.C. 1982. Significance of the presence of allophane and imogolite in Podzol B horizons for podzolization mechanisms: a review. *Soil Sci. Plant Nutr.* 28:571–578.

Farmer, V.C., and A.R. Fraser. 1979. Synthetic imogolite. In: M.M. Mortland and V.C. Farmer (eds.), *International Clay Conference, 1978.* pp. 547–553. Elsevier Scientific Publishing Co., Amsterdam. 662 pp.

Farmer, V.C., A.R. Fraser, and J.M. Tait. 1977. Synthesis of imogolite: a tubular aluminum silicate polymer. *J. Chem. Soc. Chem. Comm.* 13:462–463.

Farmer, V.C., J.D. Russell, and M.L. Berrow. 1980. Imogolite and proto-imogolite allophane in Spodic horizons: evidence for a mobile aluminum silicate complex in Podzol. *J. Soil. Sci.* 31:673–684.

Flach, K.W. 1969. The use of the 7th approximation for the classification of soils from volcanic ash. In: *Panel on Volcanic Ash Soils in Latin America.* Turrialba, Costa Rica. A.7.1–A.7.18.

Galindo, G.G., and F.T. Bingham. 1977. Homovalent and heterovalent cation exchange equilibria in soils with variable surface charge. *Soil Sci. Soc. Am. J.* 41:883–886.

Gibbs, H. 1968. *Volcanic-ash soils in New Zealand.* New Zealand Department of Scientific and Industrial Research. Information Series No. 65. 39 pp.

Gonzales-Batista, A., J.M. Hernandez-Moreno, E. Fernandez-Caldas, and A.J. Herbillon. 1982. Influence of silica content on the surface charge characteristics of allophanic clays. *Clays Clay Min.* 30:103–110.

Gradwell, M.W. 1974. The available-water capacities of some southern and central zonal soils of New Zealand. *New Zealand J. Agr. Res.* 17:465–478.

Gradwell, M.W. 1976. Available-water capacities of some intrazonal soils of New Zealand. *New Zealand J. Agr. Res.* 19:69–78.

Greenland, D.J., and J.P. Quirk. 1962. Surface areas of soil colloids. *Trans. Comm. IV and V. Intl. Soil Sci. Soc., New Zealand.* pp. 79–87.

Gunjigake, N., and K. Wada. 1981. Effects of phosphorus concentration and pH on phosphate retention by active aluminum and iron of Ando soils. *Soil Sci.* 132:347–352.

Henmi, T., and K. Wada. 1976. Morphology and composition of allophane. *Am. Mineral.* 61:379–390.

Higashi, T., and K. Wada. 1977. Size fractionation, dissolution analysis, and infrared spectroscopy of humus complexes in Ando soils. *J. Soil Sci.* 28:653–663.

Hunsaker, V.E., and P.F. Pratt. 1971. Calcium-magnesium exchange equilibria in soils. *Soil Sci. Soc. Am. Proc.* 35:151–152.

Inoko, A., and M. Tamai. 1976. Studies on soil humus with special reference to its molecular weight systems. *Bull. Nat. Inst. Agr. Sci. Japan* B28:119–182. (In Japanese)

Inoue, K. 1981. Implications of eolian dusts to 14-Å minerals in the volcanic ash soils in Japan. *Pedologist* 25:97–118. (In Japanese)

Inoue, M., S. Takeishi, H. Miyake, N. Watanabe, R. Morimoto, K. Yamaoka, and M. Sakai. 1980. Application of cartap granules into the soil for rice seedling box. II. *J. Takeda Res. Lab.* 39:21–27. (In Japanese)

Ishizawa, S., and K. Toyoda. 1964. Microflora of Japanese soils. *Bull. Nat. Inst. Agr. Sci. Japan* B14:203–284. (In Japanese)

Kato, Y. 1970. A model for amorphous matters of humic soils in Japan—a preliminary report. *Pedologist* 14:16–21. (In Japanese)

Kirkman, J.H. 1977. Possible structure of halloysite disks and cylinders observed in some New Zealand rhyolitic tephras. *Clay Min.* 12:199–216.

Kita, D., R. Nakata, and M. Harada. 1969. Geochemical study on ashy soil in Kyushu district and its lime stabilization. *J. Clay Sci. Soc. Japan* 9:28–40. (In Japanese)

Kitajima, S. 1983. Changes of soil loss and erosion patterns due to the grassland reclamation on slopes covered with volcanic ash soil. *Bull. Kyushu Nat. Agr. Exp. Sta.* 23:205–234. (In Japanese)

Knight, B.A.G., and T.E. Tomlinson. 1967. The interaction of paraquat (1:1'-dimethyl 4:4' dipyridylium dichloride) with mineral soils. *J. Soil Sci.* 18:233–243.

Kobayashi, Y. 1981. Studies of cobalt deficient grasslands for ruminants in soils of volcanic ash origin. *Jap. J. Soil Sci. Plant Nutr.* 52:394–400. (In Japanese)

Kobo, K., and Y. Oba. 1974a. Genesis and characteristics of volcanic ash soil in Japan. Pt. 7. *J. Sci. Soil Man. Japan* 45:227–233. (In Japanese)

Kobo, K., and Y. Oba. 1974b. Genesis and characteristics of volcanic ash soil in Japan. Pt. 8. *J. Sci. Soil Man. Japan* 45:293–297.

Kubota, T. 1976. Surface chemical properties of volcanic ash soil—especially on phenomenon and mechanisms of irreversible aggregation of the soil by drying. *Bull. Nat. Inst. Agr. Sci. Japan* B28:1–74. (In Japanese)

Kuroboku Soken. 1983. *International correlation of kuroboku soils (volcanic ash soils) and related soils.* Faculty of Agriculture, Kyushu University, Fukuoka. 61 pp. (In Japanese)

Leamy, M.L. 1983. Proposed revision of the Andisol proposal. *ICOMAND Circular Letter* 5:3–29.

Machida, H., and F. Arai. 1978. Akahoya ash—a Holocene wide-spread tephra erupted from the Kikai Caldera, South Kyushu, Japan. *Quat. Res.* 17:143–163. (In Japanese)

Maeda, T., and K. Soma. 1979. Soil water characteristics of organo-volcanic ash soils (Kuroboku soil). *Trans. Japan Soc. Irrig. Drain. Reclam. Eng.* 84:61–67. (In Japanese)

Maeda, T., K. Soma, and B.P. Warkentin. 1983. Physical and engineering characteristics of volcanic ash soils in Japan compared with those in other countries. *Irrig. Eng. Rur. Plan.* 3:16–31.

Maeda, T., H. Takenaka, and B.P. Warkentin. 1977. Physical properties of allophane soils. *Adv. Agron.* 29:229–264.

Martin, A.E., and R. Reeve. 1958. Chemical studies of Podzolic alluvial horizons. Pt. 3. *J. Soil Sci.* 9:89–100.

McKeague, J.A., and J.H. Day. 1966. Dithionite- and oxalate-extractable Fe and Al as aids in differentiating various classes of soils. *Can. J. Soil Sci.* 46:13–22.

Miki, N., Y. Kondo, and S. Tamura. 1975. Trace elements in volcanic ash soils distributed in Eastern Hokkaido. *Res. Bull. Obihiro Zootech. Univ.* 9:547–587. (In Japanese)

Misono, S. 1964. Soil moisture. In: *Volcanic ash soils in Japan.* pp. 75–79. Ministry of Agriculture and Forestry Japanese Government, Tokyo. 211 pp.

Mizota, C. 1981. Geomorphological relationships of copper contents of Ando soils deficient in copper, Abashiri County, Eastern Hokkaido. *J. Sci. Soil Man. Japan* 52:99–106. (In Japanese)

Mizota, C. 1982. Tropospheric origin of quartz in Ando soils and Red-Yellow soils on basalts, Japan. *Soil Sci. Plant Nutr.* 28:517–522.

Mizota, C. 1983. Eolian origin of the micaceous minerals in an Ando soil from Kitakami, Japan. *Soil Sci. Plant Nutr.* 29:379–382.

Mizota, C., and K. Wada. 1980. Implications of clay mineralogy to the weathering and chemistry of Ap horizons of Ando soils. *Geoderma* 23:49–63.

Mizota, C., M.A. Carrasco, and K. Wada. 1982. Clay mineralogy and some chemical properties of Ap horizons of Ando soils used for paddy rice in Japan. *Geoderma* 27:225–237.

Nakai, M., and N. Yoshinaga. 1980. Fibrous goethite in some soils from Japan and Scotland. *Geoderma* 24:143–158.

Nomoto, K., K. Araki, M. Ishikawa, Y. Kamada, Y. Kosegawa, and S. Yoshioka. 1955. Studies on the chemical properties of upland field soil in Tohoku district. *Bull. Tohoku Nat. Agr. Exp. Sta.* 21:30–144. (In Japanese)

Okamura, Y., and K. Wada. 1983. Electric charge characteristics of horizons of Ando (B) and Red-Yellow B soils and weathered pumices. *J. Soil Sci.* 34:287–295.

Okamura, Y. and K. Wada. 1984. Ammonium-calcium exchange equilibria in soils and weathered pumices that differ in cation-exchange materials. *J. Soil Sci.* 35:387–396.

Ono, S., and Y. Uchida. 1979. Adsorption of ammonium phosphate salt by soils. *J. Sci. Soil Man. Japan* 50:555–560. (In Japanese)

Orbell, G.E., R.C. Parfitt, M. Russell, and S.M. Robertson. 1981. Some properties of a sequence of Andosols under different rainfall. *Soils with Variable Charge, Programme and Abstracts.* pp. 167–168.

Parfitt, R.L. 1980. Chemical properties of variable charge soils. In: B.K.G. Theng (ed.), *Soils with variable charge.* pp. 167–194. New Zealand Society of Soil Science, Lower Hutt. 488 pp.

Parfitt, R.L., and T. Henmi. 1980. Structure of some allophanes from New Zealand. *Clays Clay Min.* 28:285–294.

Rajan, S.S.S. 1975. Mechanism of phosphate adsorption by allophanic clays. *New Zealand J. Sci.* 18:93–101.

Russell, M., R.L. Parfitt, and G.G.C. Claridge. 1981. Estimation of the amounts of allophane and other materials in the clay fraction of an Egmont loam profile and other volcanic ash soils, New Zealand. *Austr. J. Soil Res.* 19:185–195.

Saigusa, M., S. Shoji, and T. Takahashi. 1980. Plant root growth in acid Andosols from northeastern Japan. 2. Exchange acidity $Y_1$ as a realistic measure of aluminum toxicity potential. *Soil Sci.* 130:242–250.

Schalscha, E.B., P.F. Pratt, and L. De Andrade. 1975. Potassium–calcium exchange equilibria in volcanic ash soils. *Soil Sci. Soc. Am. Proc.* 39:1069–1072.

Schalscha, E.B., P.F. Pratt, and D. Soto. 1974. Effect of phosphate adsorption on the cation exchange capacity of volcanic ash soils. *Soil Sci. Soc. Am. Proc.* 38:539–540.

Schalscha, E.B., P.F. Pratt, T. Kinjo, and J. Amar. 1972. Effect of phosphate salts as saturating solutions in cation exchange capacity determinations. *Soil Sci. Soc. Am. Proc.* 36:912–914.

Seino, K., K. Yamashita, T. Motomatsu, and M. Motooka. 1976. The growth pattern of rice plant in several main paddy soils in the southern district of Japan. *Bull. Kyushu Agr. Exp. Sta.* 18:133–156. (In Japanese)

Sherman, D., and L.D. Swindale. 1964. Hawaiian soils from volcanic ash. In: *World soil resources reports 14.* pp. 36–49. FAO/UNESCO. 169 pp.

Shoji, S. 1983. Mineralogical properties of volcanic ash soils. In: N. Yoshinaga (ed.), *Volcanic ash soil—Genesis, Properties, Classification.* pp. 31–72. Hakuyusha, Tokyo. 204 pp. (In Japanese)

Shoji, S., and J. Masui. 1971. Opaline silica of recent volcanic ash soils in Japan. *J. Soil Sci.* 22:101–108.

Shoji, S., and T. Ono. 1978. Physical and chemical properties and clay mineralogy of Andosols from Kitakami, Japan. *Soil Sci.* 126:297–312.

Shoji, S., and I. Yamada. 1977. Soil mineralogy and fertility of Ando soils in Japan. *Proc. Intl. Sem. Soil. Env. Fert. Man. Int. Agr., Soc. Sci. Soil Man. Japan.* pp. 96–102. Tokyo.

Shoji, S., Y. Fujiwara, I. Yamada, and M. Saigusa. 1982. Chemistry and clay mineralogy of Ando soils, Brown forest soils, and Podzolic soils formed from recent Towada ashes, northeastern Japan. *Soil Sci.*. 133:69–86.

Sixth International Soil Classification workshop. 1984. *Tour guide. Part 1. Chile and Part 2. Ecuador.* Soil Management Support Services, Washington, DC.

Smalley, I.J., C.W. Ross, and J.S. Whitton. 1980. Clays from New Zealand support the inactive particle theory of soil sensitivity. *Nature* 288:576–577.

Soil Survey Staff. 1975. Soil Taxonomy. A Basic System for Making and Interpreting Soil Surveys. Dept. Agr. Handbook No. 436. Washington, DC.

Stewart, R.B., V.E. Neall, J.A. Pollok, and J.K. Syers. 1977. Parent material stratigraphy of an Egmont loam profile, Taranaki, New Zealand. *Austr. J. Soil Res.* 15:177–190.

Tait, J.M., N. Yoshinaga, and B.D. Mitchell. 1978. The occurrence of imogolite in some Scottish soils. *Soil Sci. Plant Nutr.* 24:145–151.

Tan, K.H. 1964. The Andosols in Indonesia. In: *World soil resource report 14.* pp. 30–35. FAO/UNESCO. 169 p.

Theng, B.K.G. 1972. Adsorption of ammonium and some primary *n*-alkyl-ammonium cations by soil allophane. *Nature* 238:150–151.

Theng, B.K.G., D.J. Greenland, and J.P. Quirk. 1967. Adsorption of alkyl-ammonium cations by montmorillonite. *Clay Min.* 7:1–17.

Theng, B.K.G., M. Russell, G.J. Churchman, and R.L. Parfitt. 1982. Surface properties of allophane, halloysite, and imogolite. *Clays Clay Min.* 30:143–149.

Thorp, J., and G.D. Smith. 1949. Higher categories of soil classification: order, suborder, and great soil groups. *Soil Sci.* 67:117–126.

Tokashiki, Y., and K. Wada. 1975. Weathering implications of the mineralogy of clay fractions of two Ando soils, Kyushu. *Geoderma* 14:47–62.

Tokudome, S., and I. Kanno. 1968. Nature of the humus of some Japanese soils. *Trans. 9th Intl. Congr. Soil Sci.* 3:163–173.

Tsutsumi, M., K. Ohira, and A. Fujiwara. 1968. Copper deficiency in humus rich volcanic ash soil, Pt.4. *J. Sci. Soil Man. Japan* 39:131–136 (In Japanese).

Wada, K. 1977. Allophane and imogolite. In: J.B. Dixon and S.B. Weed (eds.), *Minerals in soil environments.* pp. 603–638. Soil Science Society of America, Madison. 948 pp.

Wada, K. 1980. Mineralogical characteristics of Andisols. In: B.K.G. Theng (ed.), *Soils with variable change.* pp. 87–107. New Zealand Society of Soil Science, Lower Hutt. 448 pp.

Wada, K. 1981. Ion exchange and adsorption by soil clays. In: T. Fujisawa and N. Yoshinaga (eds.), *Adsorption phenomena in soil—Fundamental and application.* pp. 5–57. Hakuyusha, Tokyo. 160 pp. (In Japanese)

Wada, K., and A. Abd-Elfattah. 1978. Characterization of zinc adsorption sites in two mineral soils. *Soil Sci. Plant Nutr.* 24:417–426.

Wada, K., and S. Aomine. 1973. Soil development on volcanic materials during the Quaternary. *Soil Sci.* 116:170–177.

Wada, K., and T. Higashi. 1976. The categories of aluminum- and iron-humus complexes in Ando soils determined by selective dissolution. *J. Soil Sci.* 27:357–368.

Wada, K. and Y. Kakuto, 1985. Spot test with toluidine blue for allophane and imogolite. *Soil Sci. Soc. Am. J.*, in press.

Wada, K., and Y. Okamura. 1980. Electric charge characteristics of Ando $A_1$ and buried $A_1$ horizon soils. *J. Soil Sci.* 31:307–314.

Wada, K., and Y. Tange. 1984. Interaction of methyl- and ethyl-ammonium ions and piperidinium ions with soils. *Soil Sci.* 137:315–323.

Wada, K., H. Gondo, and S. I. Wada. 1982. An incipient form of halloysite in a Kuroboku paddy soil. *Abst. 1982 Mtng., Japanese Soc. Soil Plant Nutr.* 28:32. (In Japanese)

Wada, K., T. Henmi, N. Yoshinaga, and S.H. Patterson. 1972. Imogolite and allophane formed in saprolite of basalt on Maui, Hawaii. *Clays Clay Min.* 20:375–380.

Wada, S.I., and C. Mizota. 1982. Iron-rich halloysite (10 Å) with crumpled lamellar morphology from Hokkaido, Japan. *Clays Clay Min.* 30:315–317.

Wada, S.I., and A. Nagasato. 1983. Formation of silica microplates by freezing dilute silicic acid solutions. *Soil Sci. Plant Nutr.* 29:93–95.

Wada, S.I., and K. Wada. 1977. Density and structure of allophane. *Clay Min.* 12:289–298.

Wada, S.I., A. Eto, and K. Wada. 1979. Synthetic allophane and imogolite. *J. Soil Sci.* 30:347–355.

Wesley, L.D. 1973. Some basic engineering properties of halloysite and allophane clays in Java, Indonesia. *Geotechnique* 23:471–494.

Wielemaker, W.G., and T. Wakatsuki. 1984. Properties, weathering and classification of some soils formed in peralkaline volcanic ash in Kenya. *Geoderma* 32:21–44.

Wright, A.C.S. 1964. The "Andosols" or "Humic Allophane" soils of South America. In: *World soil resources report* 14. pp. 9–22. FAO/UNESCO. 169 pp.

Yagi, H., M. Takami, N. Tanaka, and T. Kubo. 1983. Properties of forest soils derived from volcanic ash on Mt. Fuji and Mt. Amagi. *Nippon Ringakukai Taikai Koenshu* 94:729–746. (In Japanese)

Yamada, I., and S. Shoji. 1982. Retention of potassium by volcanic glasses of the topsoils of Andosols in Tohoku, Japan. *Soil Sci.* 133:208–212.

Yamada, S. 1968. *Soil genesis, classification, survey and their application with emphasis on volcanic ash soils.* Yokendo, Tokyo. (In Japanese)

Yonebayashi, K. 1976. Studies on organo-mineral complexes in soils. *Sci. Rep. Kyoto Pref. Univ. Agr.* 28:121–171.

# Index